# POLE OF INSPIRATION
## MATH, SCIENCE AND TRADITION

## by Miguel Iradier

ISBN: 9798638646547

# Índice

# POLE OF INSPIRATION
## MATH, SCIENCE AND TRADITION

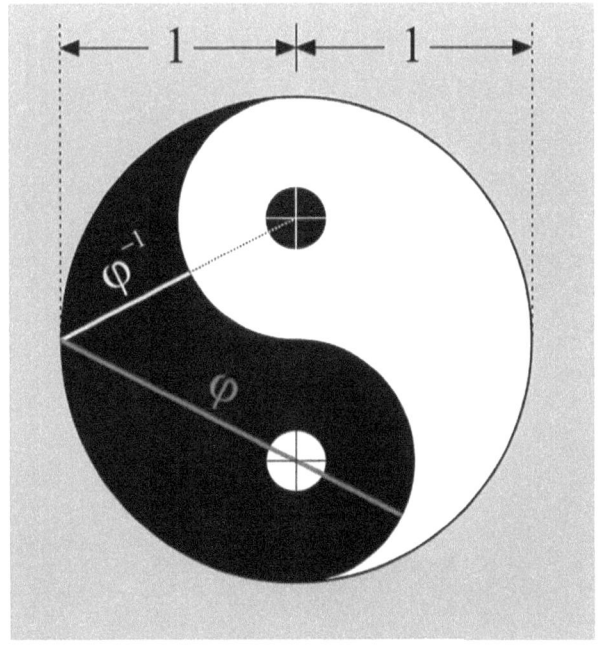

## by Miguel Iradier

# 1. The spark and the thread

Those who like simple problems can try to demonstrate this relationship before moving on. It's insultingly easy:

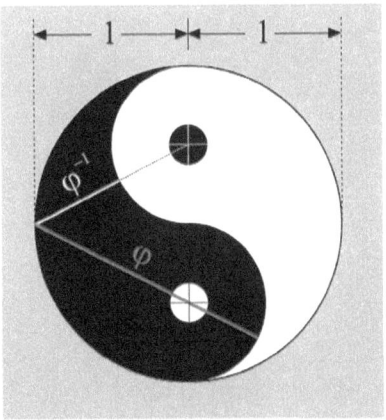

$$\varphi = 1/\varphi + 1 = \varphi^{-1} + 1 = 1/\varphi^{-1}$$

We owe this fortunate discovery to John Arioni. The elementary demonstration, along with other unexpected relationships, is on the site *Cut the knot* [1]. The number $\varphi$ is, naturally, the golden ratio $(1+\sqrt{5})/2$, in decimal figures $1.6180339887...$, and $\varphi^{-1}$ is the reciprocal, $0.6180339887...$, and since its infinite decimal places can be calculated by means of the simplest continuous fraction, here we will also call it the *continuous ratio* or *continuous propor-tion*, because of its unique role as mediator between discrete and continuous aspects of nature and mathematics.

This looks like the typical casual association of recreational math pages. One can get φ in many different ways with circles, but to my knowledge this is by far the most elementary of all, being the radius the unit of reference. In other words, this relation seems too simple and direct not to contain something important. And yet it has only recently been discovered, almost by chance.

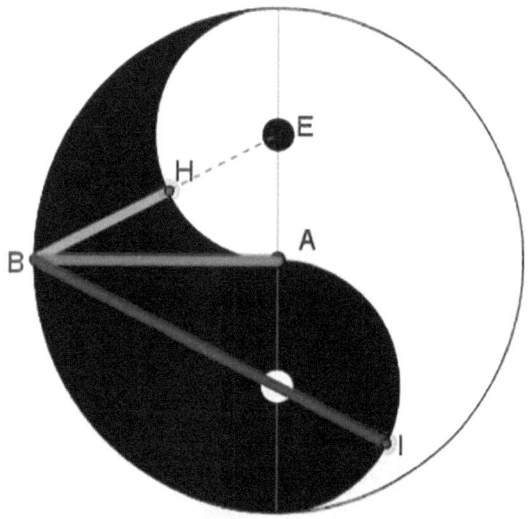

John Arioni

Since Euclid and probably much earlier, the entire history of findings on this proportion has been derived from the division of a segment "in the extreme and mean ratio", and has developed with the construction of squares and rectangles. The most immediate cases involving the circle come from the construction of the penta-gon and the pentagram, no doubt known to the Pythagoreans; but one does not need to know anything about mathematics to realize that the relationship contained in this symbol is of a much more fundamental order —just as, from the quantitative point of view, the 2 is closer to 1 than 5, or from the qualitative one, the dyad is closer to the monad than the pentad.

If the circle and its central point are the most general and comprehensive symbol of the monad or unit, we have here the most immediate and revealing proportion of reciprocity, or dynamic symmetry, presented after the division into two parts. The *Taijitu* has a double function, as a symbol of the supreme Pole, beyond duality, and as a representation of the first great polarity or duality. It is, as it were, halfway between both, and both are linked by a ternary relationship —precisely the continuous proportion.

A relation is the perception of a dual connection, while a proportion implies a third order relationship, a "perception of perception". Since at least the times of the Kepler's triangle, we have known that the golden mean articulates and conjugates in itself the three most fundamental means of mathematics: the arithmetic mean, the geometric mean, and the so-called harmonic mean between both.

We could ask ourselves what would have happened if Pythagoras had known about this correlation, which would certainly have exalted Kepler's imagination as well. It will be said that, like any other counterfactual, there is no point about it. But the question is not as much about the past that might have been as it is about the possible future. Pythagoras could hardly have been as surprised as we are, since he knew nothing about the decimal values of $\varphi$ or $\pi$. Today we know that they are two ratios running to an infinite number of decimal places, and yet they are linked exactly by the most elementary triangular relationship.

Mathematical truth is beyond time, but not its revelation and construction. This allows us to see certain things with the insight of a Geohistory, as it were, in four dimensions. There has been speculation about what would have happened if the Greeks had known and made use of the zero, and whether they might have developed modern calculus. This is very doubtful, since they would have still needed to make a series of great leaps far from their conception of the world, such as the numbering system, the zero and its positional use, the idea of derivative, and so on. The double spirals were a common motif in archaic Greece, and the arithmetical speculations of the Pythagoreans, very similar in nature to those developed by the Chinese over time; but for whatever reason the Greeks did not intertwine the two spirals into one, and, in China itself, a diagram like the one we know today did not came into existence until the end of the Ming dynasty and only after a lengthy evolution.

Which is just another example of how hard is to see the obvious. It's not so much the thing itself, but the context in which it emerges and in which it fits. Depending on how one looks at it, this can be encouraging or discouraging. In knowledge there is always a high margin to simplify, but as in so many other things, that margin depends to a large extent on knowing how to make it happen.

The *Taijitu*, the symbol of the supreme Pole, is a circle, a wave and a vortex all in one. Of course, the vortex is reduced to its minimum expression in the form of a double spiral. Characteristically, the Greeks separated their double spirals, and eventually turn them into squares, in the motifs known today as grecas. It is just another expression of their taste for statics, a bent that set the general framework for the reception of the golden mean in mathematics and art, and which has come down to us through the Renaissance.

The series of numbers that approximate infinitely the continuous proportion, known to us as Fibonacci numbers, appeared already long before in the numerical triangles consecrated in India to Mount Meru, "the mountain that surrounds the world", which is just another designation of the Pole. As it is well known, from this figure, called the Pascal's triangle in the West, a huge number of combinatory properties, scales and sequences of musical notes are derived.

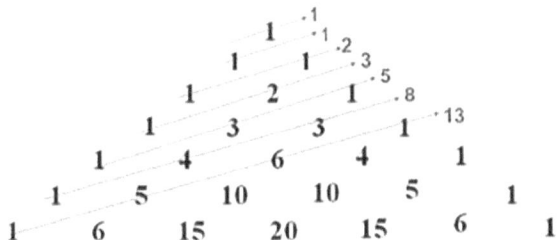

The polar triangle, known in other cultures as Khayyam's triangle or Yang Hui's triangle, is one of those "extraordinarily well connected" mathematical objects: from it one can derive the binomial expansion, the binomial and normal statistical distributions, the $\sin(x)^{n+1}/x$ transform of harmonic analysis, the matrix exponential and the exponential function, or the values of the two great gears of calculus, the constants $\pi$ and $e$. It is almost incredible that the elemental connection with the Euler number has not been discovered until 2012 by Harlan J. Brothers. Instead of adding up all the figures in each row, one only needs to extract the ratio of ratios for their product; the difference between sums and products is a motif that will emerge several times throughout this article.

The polar triangle looks like an arithmetic and "static" representation, while the *Taijitu* is like a geometric snapshot of something purely dynamic. However, the rich implications for music of this triangle, partially explored by the work of Ervin Wilson, largely circumvent the separations created by adjectives such as "static" and "dynamic". In any case, if the staircase of figures deployed in Mount Meru is an infinite progression, when we finally see the lines hidden in the circular diagram of the Pole we immediately know that it is something irreducible —the first offers us its arithmetical deployment and the second its geometric retraction.

The oldest known mention of the triangle, albeit a cryptical one, can be found in the Chandaḥśāstra of Pingala, where Mount Meru is shown as the formal archetype for metric variants in versification. It is also fair to say that the first Chinese author to deal with the polar triangle is not Yang Hui but Jia Xian (ca. 1010-1070), a strict contemporary of the philosopher and cosmologist

Zhou Dunyi (1017-1073), the first author who publicized the *Tai-jitu* diagram.

Nowadays very few people are aware that both figures are representations of the Pole. It is my conjecture that all the mathematical relationships that can be derived from the polar triangle can also be found in *Taijitu*, or at least generated from it, although under a very different aspect, and with a certain twist that possibly involves φ. Both would be a dual expression of the same unity. Mathematicians will see what is the point of this.

Between counting and measuring, between arithmetic and geometry, we have the basic areas of algebra and calculus; but there is an overwhelming evidence that the latter branches have developed in one particular direction more than in others —more in decomposition than in composition, more in addition than in multiplication, more in analysis than synthesis. So the study of the relations between this two expressions of the Pole could be full of interesting surprises and basic but not trivial results, and it poses a different orientation for mathematics.

It can be seen that the arithmetic triangle has closed links with fundamental aspects of calculus and the mathematical constant *e*, while the *Taijitu* and the constant φ lack in this respect relevant connections —hence the totally marginal character of the continuous proportion in modern science. It has been said that the latter, unlike the intimate connection with change of Euler's number, is a static relationship. However, its appearance in the extremely dynamic character of the yin-yang symbol already warns us of a general change of context.

For centuries calculus has been dissolving the relationship between geometry and change in favor of algebra and arithmetic, of not so pure numbers. Now we can turn this sandglass upside down observing what happens on the upper bulb, the lower bulb and the neck.

\*

The appearance of the golden mean between the yin and yang in a purely curvilinear fashion not only is not static but on the contrary cannot be more dynamic and functional, and indeed the *Taijitu* is the most complete expression of activity and dynamism with the minimum number of elements. The diagram also has an intrinsic organic and biological connotation, inevitably evoking cell division, which in fact is an asymmetrical process, and, at least in plant growth, often follows a sequence governed by this ratio. In other words, the context in which the continuous ratio emerges here is the true antithesis of its Greek reception that has lasted until today, and this can have far-reaching implications on our perception of this proportion.

Oleg Bodnar has developed an elegant mathematical model of plant phyllotaxis with hyperbolic golden functions in three dimensions and with coefficients of reciprocal expansion and contraction that can be seen in the great panoramic book that Alexey Stakhov dedicates to the Mathematics of Harmony [2]. It is an example of dynamic symmetry that can be perfectly combined with the great diagram of polarity, regardless of the nature of the underlying physical forces.

The presence of spiral patterns based on the continuous proportion and their numerical series in living beings does not seem mysterious. Whether in the case of a nautilus or vegetable tendrils, the logarithmic spiral —the general case- allows indefinite growth with no change of shape. Spirals and helixes seem an inevitable result of the dynamics of growth, by the constant accretion of material on what is already there. At any rate, we should ask why among all the possible proportions of the logarithmic spiral those close to this constant arise so often.

And the answer would be that the discrete approaches to the continuous proportion also have optimal properties from several points of view —and cell growth ultimately depends on the discrete process of cell division, and at higher levels of organization, on other discrete elements such as tendrils or leaves. Since the convergence of the continuous ratio is the slowest, and plants tend to fill as much room as possible, this ratio allows them to emit the greatest number of leaves in the space available.

This explanation seems, from a descriptive point of view, sufficient, and makes it unnecessary to invoke natural selection or deeper physical mechanisms. However, in addition to the basic discrete-continuos relationship, it contains implicitly a powerful link between forms generated by an axis, such as the pine cones, and the so-called "principle of maximum entropy production" of thermodynamics, which we will find later again.

Needless to say, we do not think this proportion has "the secret" to any universal canon of beauty, since surely such a canon does not even exist. However, its recurrent presence in the patterns of nature shows us different aspects of an spontaneous principle of organization, or self-organization, behind what we superficially call "design". On the other hand, the appearance of this mathematical constant, due to its very irreducible properties, in a great number of problems of optimization, maximums and minimums, and parameters with critical points allows us to connect it both naturally and functionally with human design and its search for the most efficient and elegant configurations.

The emergence of the continuous proportion in the dynamic symbol of the Pole —of the very principle- augurs a substantive change both in the contemplation of Nature and in the artificial constructions of human beings. Contemplation and construction are antagonistic activities. One goes top-down and the other bottom-up, but there is always some sort of balance between both. Contemplation allows us to free ourselves from the connections already built, and construction gets ready to fill the resulting void with new ones.

It is somewhat strange that the continuous proportion, despite its frequent presence in Nature, is so poorly connected with the two great constants of calculus, $\pi$ and $e$ —except for anecdotic incidences as the "logarithmic golden spiral", which is only a particular case of an equiangular spiral. We know that both $\pi$ and $e$ are transcendental numbers, while $\varphi$ is not, although it is indeed the "most irrational number", in the sense that it is the one with the slowest approximation by rational numbers or fractions. φ is also the simplest natural fractal.

Until now, the most direct link with trigonometric series has been through the decagon and the identities φ = 2cos 36° = 2cos (π/5). It has not been associated so far with imaginary numbers, $i$ being the other great constant of calculus, which is concurrent with the other two in Euler's formula, of which the so-called Euler identity ($e^{i\pi} = -1$) is a particular case.

The number $e$, base of the function that is its own derivative, appears naturally in rates of change, the subdivisions ad infinitum of a unit that tend to a limit or in wave mechanics. The imaginary numbers, on the other hand, so common in modern physics, appeared for the first time with the cubic equations and pop up each time additional degrees of freedom are assigned to the complex plane.

Actually, complex numbers behave exactly like two-dimensional vectors, in which the real part is the inner or scalar product or and the so-called imaginary part corresponds to the cross or vector product; so imaginary numbers can only be associated with motions, rotations and positions in space in additional dimensions, not with the physical quantities themselves.

This is easier to say than to think of, since it is even more "complex" to determine what a physical quantity or a mathematical variable can be independently of change and motion. Both to geometrically interpret the meaning of vectors and complex numbers in physics and to generalize them to any dimension a tool like geometric algebra may be used —"the algebra flowing from geometry", as Hestenes put it; but even then there is much more to geometry than we may think.

Many problems become more simple on the complex plane, or so the mathematicians say. One of them, under the pseudonym *Agno* sent in 2011 an entry to a math forum with the title "Imaginary Golden Mean", which shows a direct connection with $\pi$ and $e$ : $\Phi i = e^{\pm \pi i/3}$ [3]. Another anonymous author found this same

[17]

identity in 2016, along with similar derivations, looking for fundamental properties of an operation known as "reciprocal addition", of interest in circuits and parallel resistances calculations. As refraction is a kind of impedance, it may also have its place in optics. The relation in the polar diagram may be associated right from the start with geometric series and hypergeometric functions associated with continuous fractions, modular forms and Fibonacci series, and even with noncommutative geometry [4]. The imaginary golden ratio, in any case, reflects as in a mirror many of the qualities of its real part.

The *Taijitu* is a circle, a wave and a vortex all in one. The synthetic genius of nature is quite different from that of man, and she does not ask for unifications because not to arbitrarily separate is enough for her. Nature, as Fresnel said, does not care about analytical difficulties.

The diagram of the *Taijitu* becomes a flat section of a double spiral expanding and contracting in three dimensions, a motion that seems to give it an "extra dimension" in time. It is always a real challenge to follow the evolution of this process, both spiral and helical, within a vertical cylinder, which is but the complete representation of the *indefinite propagation of a wave motion*, the "universal spherical vortex" described by René Guenon in three short chapters of his work "The Symbolism of the Cross". The cross of which Guenon speaks is certainly a system of coordinates in the most metaphysical sense of the word; but the physical side of the subject is by no means negligible.

The propagation of a wave in space is a process as simple as it is difficult to grasp in its entirety; one need only think of Huygens' principle, the universal mode of propagation, which also underlies all quantum mechanics, and which involves continuous deformation in a homogeneous medium.

In that same year of 1931 when Guenon was writing about the evolution of the universal spherical vortex, the first work was published on what we know today as the Hopf fibration, the map of the connections between a three-dimensional sphere and a sphere in two dimensions. This enormously complex fibration is found even in a simple two-dimensional harmonic oscillator. Also in the same year, Paul Dirac conjectured the existence of that unicorn of modern physics known as the magnetic monopole, which brought the same kind of evolution into the context of quantum electrodynamics.

Peter Alexander Venis gives us in a wonderful work a completely phenomenological approach to the classification and typology of the different vortices. There is nothing mathematical here, neither advanced nor elementary, but a sequence of transformations of 5 + 5 + 2, or 7 classes of vortices with many types and countless variants that unfold from the completely undifferentiated only to return to the undifferentiated again —or to the infinity of which Venis prefers to speak. The transitions from ideal points with no extension to the apparent forms of nature seems quite arbitrary without the aid of vortices, hence their importance and universality.

Peter Alexander Venis

Venis does not deal with the mathematical and physical aspects of such a complex subject as vortices, and of course he does not apply to them the continuous proportion; on the contrary he gives us the privilege of a new fresh vision of these rich processes, in which the insight of a presocratic naturalist and the capacity for synthesis of a Chinese systematist meet together effortlessly.

Even if the Venis sequence admits variations, it presents us a morphological model of evolution that goes beyond the scope of ordinary sciences and disciplines. The author includes under the term "vortices" flow processes that may or may not have rotation, but there is a good reason for that, since this is necessary to cover key conditions of equilibrium. He also applies the theory of yin and yang in a way that is both logical and intuitive, which probably admits a fairly elementary translation to the qualitative principles of other traditions.

The study of this sequence of transformations, in which questions of acoustics and image are closely linked, should be of immediate interest in order to deepen the criteria of morphology and design even without the need to enter into further considerations.

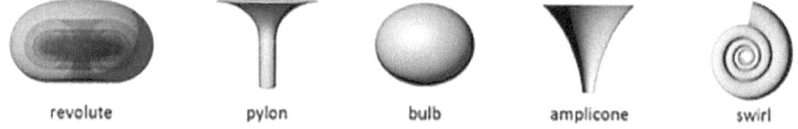

| revolute | pylon | bulb | amplicone | swirl |

Peter Alexander Venis

A metric-free description would be, precisely, the perfect counterpoint for a subject as badly affected by arbitrariness in the measurement criteria as the study of proportionality. Naturally, mathematics also has several tools essentially free of metrics, such as external differential forms, which allow the study of the physical fields with maximum elegance. Then, perhaps, the metrics that physics deals with could be used as a middle ground between both extremes.

Thus, in this search to better define the context for the appearance of the continuous proportion in the world of phenomena, we can speak of three types of basic spaces: the ametric or metric-free space, the metric spaces, and the parametric or parameter spaces.

By metric-free space we understand the different spaces that are free of metrics and the action of measurement, from the purely morphological sequence of vortices above to projective geometry or the metric independent parts of the topology or differential forms. The projective, metric-independent space is the only true space; if we sometimes speak of metric spaces it is only because of the different connections with metric spaces.

By metric spaces, we mean those of the fundamental theories in physics, not only mainstream theories but also other related, with a special emphasis on Euclidean metric space in three dimensions of our ordinary experience. They include physical constants and variables, but here we are particularly interested in theories that do not depend on dimensional constants and can be expressed in homogeneous proportions or quantities.

By parametric or parameter spaces we mean the spaces of correlations, data, and adjustable values that serve to define mathematical models, with any number of dimensions. We can also call it the algorithmic and statistical sector.

We are not going to deal here with the countless relationships that can exist between these three kinds of spaces. Suffice it to say that to get out of this labyrinth of complexity in which all sciences are already immersed, the only possible Ariadne's thread, if any, has to trace a retrograde path: from numbers to phenomena, with the emphasis on the latter and not the other way around. And we are referring to phenomena not previously limited by a metric space.

Much has been said about the distinction between "the two cultures" of sciences and the humanities, but it should be noted that, before attempting to close this by now unsurmountable gap, we should begin first by bridging the gap between the natural, descriptive sciences and a physical science that, justified by its predictions, becomes indistinguishable with the power of abstraction of mathematics while isolating itself from the rest of Nature, to which it would like to serve as foundation. Reversing this fatal trend is of the greatest importance for the human being, and all efforts in that direction are worthwhile.

## 2. Physics and the continuous proportion

We already see that there are purely mathematical reasons for the continuous proportion to appear in the designs of nature independently of causality, be it physical, chemical or biological: in fact the convenience of logarithmic growth is independent even of the form itself, as is the elementary fact of the discrete and asymmetric division of cells.

In this light, it would be an emergent property, just a parallel plane to physical causation and becoming. On the other hand, the idea of parallel planes with a merely circumstantial connection with physical reality looks odd, and in any case very distant from what the diagram of the Pole express so well —that no form or nothing apparent is free from dynamics.

The fact is that the connection between physics and the continuous proportion is very dim, to say the least. However we have important occurrences of this ratio even in the Solar System, where it is almost impossible to avoid celestial mechanics. A better understanding of the presence of the continuous proportion in nature should not ignore the framework defined by fundamental physical theories, nor what these might leave out.

We have three possible approaches with increasing degrees of risk and depth:

1. The continuous proportion in nature can be studied independently of the underlying physics as a purely mathematical question; this would be the most prudent, but somewhat limited position. The aforementioned A. Stakhov has developed an algorithmic theory of measurement based on this ratio that can be used to analyze in turn other metrological theories of cycles, continuous fractions and fractals as for example the so called Global Scaling.

2. This proportion can be studied according with views compatible with known mainstream physics; for example, as Richard Merrick has done, in a neopythagorean rereading of the collective harmonic aspects of wave mechanics, such as resonances, and in which *phi* would be a critical damping factor [7]. These ideas are totally accessible to the experiment, either in acoustics or in optics, so that they can be verified or falsified.

Merrick's idea of harmonic interference is within everyone's reach and understanding and it is not without depth. It can be naturally complemented with the holographic concept proposed by David Bohm and his distinction between the implicate and explicate order. Although Bohm's interpretation is not standard, it is compatible with experimental data. The harmonic interference theory can also be combined with mathematical theories of cycles and scales such as those mentioned.

3. Or, finally, one can consider other more classical theories that differ from the mainstream but which may provide deeper insights into the subject. Within this category, there are various degrees of disagreement with the standard theories: from just a wider understanding of thermodynamics, to in-depth revisions of classical mechanics, quantum mechanics and calculus. We could say that this third option is not that speculative, but rather divergent in the spirit and the interpretation.

Here we will focus more on the third level, which may also seem the most problematic. One could ask what is the need to question the best established physical theories in order to find a better ground for the occurrence of a constant that maybe does not require them. Furthermore, the first two levels already offer plenty of room for speculation. But this would be a very superficial way of looking at it.

We cannot delve into the presence of the continuous proportion in a symbol of perfect reciprocity ignoring the question of whether our present theories are the best exponent of continuity, homogeneity or reciprocity —and in fact they are far from it.

## 3. Two kinds of reciprocity

The *Taijitu*, the emblem of the action of the Pole with respect to the world, and of the reciprocal action with respect to the Pole, inevitably reminds us of the most universal figure in physics; we are naturally referring to the ellipse —or rather, it should be said, to the idea of the generation of an ellipse with its two foci, since here there is no eccentricity. The ellipse appears in the orbits of the planets no less than in the atomic orbits of the electrons, and in the study of the refractive properties of light it gives rise to a whole field of analysis, ellipsometry. Kepler's old problem has scale invariance, and plays a determining role in all our knowledge of physics from the Planck constant to the furthest galaxies.

In physics and mechanics, the principle of reciprocity par excellence is Newton's third principle of action and reaction, which is at the base of all our ideas about energy conservation and allows us to "interrogate" forces when we are obliged to assume the constancy or proportionality of other quantities. The third principle does not speak of two different forces but of two different sides of the same force.

Now, the story of the third principle is curious, because we are forced to think that Newton established it as the keystone of his system to tie up the loose ends of celestial mechanics —particularly in Kepler's problem- rather than for down-to-earth mechanics based on direct contact between bodies. The third principle allows us to define a closed system, and closed systems have been the given for all fundamental physics since then —however, it is precisely in celestial orbits, such as that of the Earth around the Sun, that this principle can be least verified, since the central body is not in the center, but in one of the foci. The force designated by the vectors would have to act on the void, where there is no matter.

Since the very first moment it was argued in the continent that Newton's theory was more an exercise in geometry than in physics, although the truth is that, if physics and vectors were good for something, the first thing that failed was geometry. That is, if we assume that forces act from and on centers of mass, instead of on mere mathematical points. But, despite what intuition tells us —that an asymmetric ellipse can only result from a variable force, or from a simultaneous generation from the two foci-, the desire to expand the domain of calculus prevailed over anything else.

In fact the issue has remained so ambiguous that attempts have always been made to rationalize it with different arguments, either the system's barycenter, or the variation in orbital velocity, or the initial conditions of the system. But none of them separately, nor the combination of the three, allows to solve the issue satisfactorily.

Since no one wants to think that the vectors are subjected to quantitative easing, and they lengthen and shorten at convenience, or that the planet accelerates and brakes on its own as a self-propelled rocket, in order to keep the orbit closed, physicists finally came to accept the combination in one quantity of the variable orbital speed and innate motion. But what happens is that if the centripetal force counteracts the orbital velocity, and this orbital velocity is variable despite the fact that the innate motion is invariable, the orbital velocity is in fact already a result of the interaction between the centripetal and the innate force, and then the centripetal force is also acting on itself. Therefore, the other options being ruled out, what we have is a case of *feedback or self-interaction* of the whole system.

So it must be said that the claim that Newton's theory explains the shape of the ellipses is at best a pedagogical resource. However, this swift pedagogy has made us forget that our so called laws do not determine or "predict" the phenomena we observe, but try to fit them at most. Understanding the difference would help us to find our place in the overall picture.

The reciprocity of Newton's third principle is simply a change of sign: the centrifugal force must be matched by an opposing force of equal magnitude. But the most elementary reciprocity of physics and calculus is that of the inverse product, as already expressed by the formula of velocity, ($v = d/t$), which is the distance divided by time. In this very basic sense, those who have pointed out that velocity is the primary fact and phenomenon of physics, from which time and space are derived, are absolutely right.

The first attempt to derive the laws of dynamics from the primary fact of velocity is due to Gauss, around 1835, when he proposed a law of electric force based not only on distance but also on relative velocities. The argument was that laws such as Newton's or Coulomb's were laws of statics rather than of dynamics. His disciple Weber refined the formula between 1846 and 1848 by including relative accelerations and a definition of potential —a retarded potential, in fact.

Weber's electrodynamic force is the first case of a complete dynamic formula in which all quantities are strictly homogeneous or proportional [8]. Such formulas seemed to be exclusive to Archimedes' statics, or Hooke's elasticity law in its original form. In fact, although it is an specific formula for electric charges and not a field equation, it allows to derive Maxwell's equations and the electromagnetic fields as a particular case, simply integrating over volume.

The logic of Weber's law could be applied equally to gravity, and in fact Gerber used it to calculate the precession of Mercury's orbit in 1898, seventeen years before the calculations of General Relativity. As is well-known, General Relativity aspired to include the so-called "Mach principle", although in the end it did not succeed; but Weber's law was entirely compatible with that principle in addition to explicitly using homogeneous quantities, well before Mach wrote about these issues.

It has been said that Gerber's argument and equation was "merely empirical", but in any other era not having to create ad hoc postulates would have been seen as the best virtue. In any case, if the new proportional law was used to calculate a tiny secular divergence, and not for the generic ellipse, it was for the simple reason that in a single orbital cycle there was nothing to calculate for either the old or the new theory.

Weber's purely relational formula cannot "explain" the ellipse either, since force and potential are simply derived from motion —but at least there is nothing *unphysical* in the situation, and the fulfillment of the third principle is guaranteed while permitting a deeper meaning.

Ironically, as this new law changed the prevailing idea of central forces, understood with a string attached, Helmholtz and Maxwell blamed Weber's law for not complying with energy conservation, although finally in 1871 Weber showed that it did so on the condition that the motion was cyclical —which in this issue was already the basic requirement for Newtonian or Lagrangian mechanics too. Conservation is a global property, not a local one, but the same was true for the orbits described in the Principia, not less than those of Lagrange. Strictly speaking there is no local conservation of forces that can make physical sense. Newton himself used the analogy of a slingshot, following Descartes' example, when he spoke of the centrifugal motion, but nowhere in his definitions is there any talk that the central forces should be understood as if connected by a string. However, posterity took the simile at face value.

Why claim that there is in any case feedback, self-interaction? Because all gauge fields, characterized by the invariance of the Lagrangian under transformations, are equivalent to a non-trivial feedback between force and potential —the eternal "information problem", namely how does the Moon know where the Sun is and how does it "know" its mass to behave as it does.

Indeed, if the Lagrangian of a system —the difference between kinetic and potential energy - has a certain value and is not equal to zero, this is equivalent to say that action-reaction is never immediately fulfilled. However, we use to assume that Newton's third principle is immediate and takes place automatically and simultaneously, without mediation of any time sequence, and the same simultaneity is assumed in General Relativity. The presence of a retarded potential indicates at least the existence of a sequence or mechanism, even if we can not say anything else about it.

This shows us that additive and multiplicative reciprocity are notoriously different; and the one shown by the continuous proportion in the diagram of the Pole includes the second kind. The first is purely external and the second is internal to the order considered.

All the misunderstandings about what is mechanics come from here. And the essential difference between a mechanical system in the trivial sense and an ordered or self-organized system lies precisely at this point.

At the time it was believed that Hertz's experiments confirmed Maxwell's equations and disproved Weber's, but that is another misunderstanding because if Weber's law —which was the first to introduce the factor of the speed of light- did not predict electromagnetic waves, it did not exclude them either. It simply ignored them. On the other hand, some perceptive observers have noted that the only thing Hertz demonstrated was the reality of the action at a distance, not of waves, but that is another story.

As a counterpoint, it is worth remembering another fact that shows, among other things, that Weber had not fallen behind his time. Between the 1850s and 1870s he developed a stable model of the atom with elliptical orbits —many decades before Bohr proposed his model of the circular atom, without the need to postulate special forces for the nucleus.

Weber's relational dynamics shows another aspect that may seem exotic in the light of the present theories: according to its equations, when two positive charges approach a critical distance, they produce a net attractive force, rather than a repulsive one. But is not the very idea of an elementary charge exotic in the first place, or should we just say a mere convention? In any case, this fits very well with the *Taijitu* diagram, in which a polarized force can potentially become its opposite. Without this spontaneous reversal, hardly we could speak of truly alive forces and potentials.

## 4. The apple and the dragon

To my knowledge, Nikolay Noskov was the first to appreciate, in the 1990s, that Weber's dynamics was so far the only one that allowed for a physical account of the shape of the ellipses, even if it did not pretend to give a "mechanical explanation" for them. In this respect, Noskov particularly insisted on associating the retarded potentials with longitudinal vibrations of the moving bodies in order to give a content to the conservation, merely formal in Weber, of energy; he also insisted that their occurrence permeated all types of natural phenomena, from the stability of atoms and their nuclei, to orbital elliptical motion, sound, light, electromagnetism, the flow of water or gusts of wind [9].

Despite the misunderstandings on the subject, these longitudinal waves are not incompatible with known physics, and Noskov recalled that the same Schrödinger wave equation is a mixture of different equations that describe waves in a medium and waves within the moving body —and the same thing happened from the start with Maxwell's "electromagnetic waves", which even from the most classical point of view cannot be anything other than a statistical average between what occurs in portions of space and matter.

Noskov noticed that the behavior of forces and potentials in Weber's law involved a sort of feedback, although he does not seem to recognize that this is already the case for all gauge theories, and finally even for Newtonian celestial mechanics itself, although in all these instances it is presented in disguise. Atoms would be definitely dumb without this ability to adjust embedded into the very idea of the field.

Let us now return to the continuous proportion. Miles Williams Mathis wonders how it is that, given the equality $\Phi^2 + \Phi = 1$, *phi* has not been related to the most elementary inverse-square laws of physics; moreover, he wonders how it is that it has not been associated with the sphere itself, being so evident that the surface of a sphere also decreases to the square [10].

It could be argued that the Fibonacci series does not square, but the factor $\Phi$ does, as can be easily seen in the successive squares of the golden spiral $(1, 1/\Phi, 1/\Phi^2, 1/\Phi^3 ...)$ or in its expression as a continuous square root. Mathis is not confusing the inverse square with the square root, but is talking about a scale factor between two hypothetical subfields one into the other.

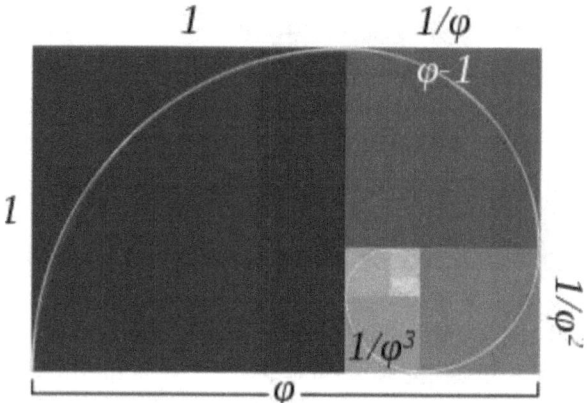

Miles Mathis: *More on the Golden Ratio and Fibonacci Series*

$$\varphi = \sqrt{1 + \sqrt{1 + \sqrt{1 + \sqrt{1 + \cdots}}}}.$$

[34]

Mathis may be right in insisting that the presence of *phi* must also have an underlying physical cause; the only problem is that modern physics ignores and completely denies a scale relationship between charge and gravity, indeed light and gravity, as he is proposing. However, the origin of his correlation lies in the same Kepler's problem, in which he wants to see a joint action of two different fields, the second one based not only on the inverse square of the distance but also on an inverse law of the fourth power ($1/r^4$) with a product of density by volume, instead of the usual formula of masses.

Now, Mathis is the first to specifically point out the conflation of orbital velocity and innate motion in Newton, interpreting the Lagrangian as the disguised product of two fields, of opposite attractive and repulsive effects, whose relative proportion or intensity is a function of scale and density [11].

The inclusion of density would have to be fundamental in a truly Archimedean relational physics, which brings us back to the issue of waves and spirals. Spirals are a common occurrence in astronomy, galaxies being their most apparent manifestation; these galaxies have been described in terms of density waves.

Many have noticed a logarithmic spiral with $\Phi$ as a key also in the Solar System and the distribution of its planets. As in the case of the so-called "law", or rather rule of Titus-Bode, the existence of a non-random order seems quite evident, but the adjustment with the known values is somewhat arbitrary.

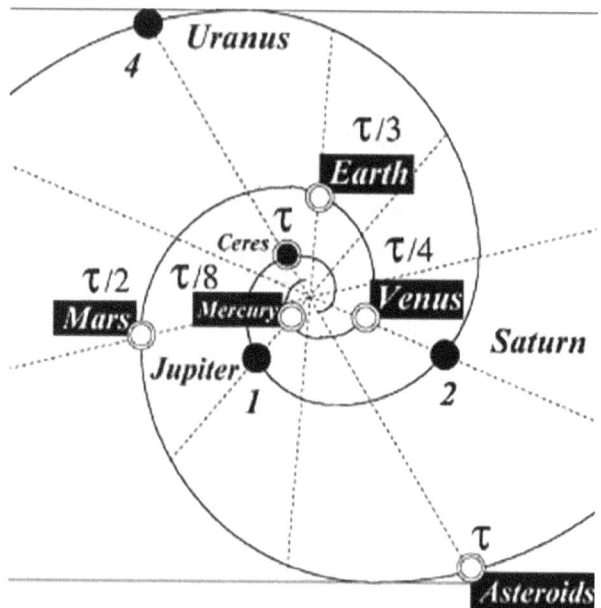

Jan Boeyens: Commensurability in the Solar System

It goes without saying that the ellipse is the transformation of the circle when its centre is divided into two foci; although from the other point of view it can be said, and this is not unimportant, that the circle is only the limit case of the first one. Expanding on Kepler's problem, although in a different light, Nicolae Mazilu refers us to Newton's theorem of revolving orbits. Newton had already carefully considered the case of forces decreasing to the cube of distance, and in this hypothetical case the bodies describe orbits with logarithmic spiral shapes, which of course no one has observed.

However, E. B. Wilson's works of 1919 and 1924 showed that the stable electron orbits in the atom were not ellipses but logarithmic spirals; only that the force involved here is not the Coulomb force, but a transition force between two different elliptical orbits. The later solution of the problem has buried in oblivion a model that was also consistent. And as for all applications of conic sections to physics, here too we find that signature of change in the potential, the shift in phase or plane of polarization known as the geometrical phase, discovered by Pancharatnam and so successfully generalized to quantum mechanics by Michael Berry [12] .

Various studies recount that the distribution of the planets in the Solar System follows the pattern of a logarithmic golden spiral with an accuracy of more than 97 percent, which may increase if the sidereal years and synodic periods of the system as a whole are taken into account [13]. For Hartmut Müller, the proximity is simply due to the closeness of *phi* to the value of √e, which is 1.648. According to other counts not verified by myself, the average distance between consecutive planets from the Sun to Pluto, taking the distance between the two previous ones as a unit, is just 1.618. If the last planet is discarded, the average deviates widely, which gives an idea of the fragility of these calibrations.

It has often been said that the perceptible harmony in the Solar System is not possible without some feedback mechanism, while the Newtonian approach simply combines a force at a distance with trajectories like cannonballs —a cannonball theory of everything- dependent on external forces or collisions. However, we have already seen that even in the Newtonian case a self-interaction is masked by merging innate motion and orbital velocity into one.

Newtonian celestial mechanics gave way to a more abstract version, Lagrangian mechanics, to avoid this mess; the difference between the kinetic energy and the potential begs the question to the so-called "initial conditions", but these are nothing but Newton's innate motion… at any rate, the case is that this average difference of the Lagrangian and the average eccentricity of the orbits is of the same order of magnitude than the deviations of the distribution of the solar system obtained by the logarithmic golden spiral. Thus, one can take the Lagrangian density of the entire system and its averages and see how the planets with their orbits nestle in.

It seems that scientific publications no longer admit studies on planetary distribution, since, having no underlying physics, they are relegated to the limbo of numerological speculation. However, the Lagrangian routinely used in celestial mechanics is also nothing more than a pure mathematical analogy, and exists only to blur differences of the same order of magnitude. Suffice it to admit this to realize that both issues are not on different grounds —maybe they are not even two different things.

To admit this is also to admit that gravity itself is an adjustment force that depends on the environment and not a universal constant, but this is something that was already implicit in Weber's relational mechanics.

Mathis' theory is more specific in that it regards G as a transformation between two radii. Not concerned with fitting his own notions of the physics underlying the Golden Section into the spiral of the Solar System, he deals in detail with Bode's Law in a much simpler way based on a series based on $\sqrt{2}$. He also includes naturally in it the optical equivalence, the neglected fact that many planets look of the same size from the Sun, just as many satellites look the same size as the Sun seen from their respective planets. So this is not a punctual coincidence [13]. The optical equivalence would be the final wink that Nature gives us to see who is more blind, she or we.

And since it looks like a typical fancy of Nature, let us allow ourselves a bit of numerological fun. The optical equivalence that is shown in the total eclipses is an angular or projective relationship (with an approximate value of 1/720 of the celestial sphere) in accordance with the number 108, so important in different traditions, here entailing the number of solar diameters between the Sun and the Earth, the number of terrestrial diameters in the diameter of the Sun and the number of lunar diameters that separate the Moon from the Earth.

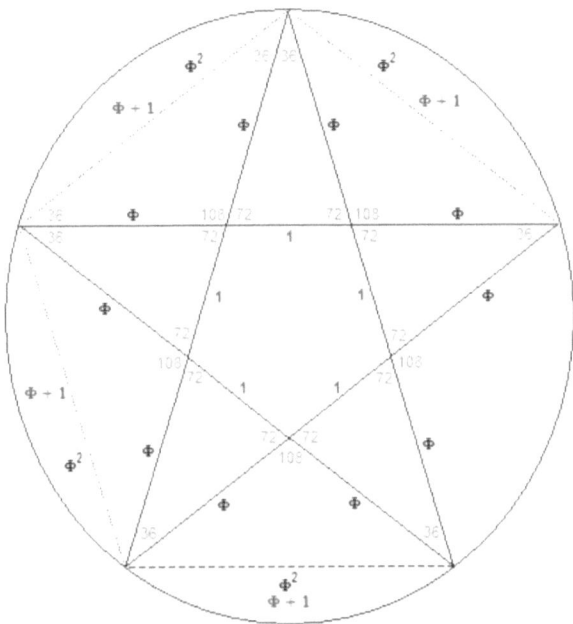

Dimensions of the pentagram and its relation to PHI - (by: Birol Koc) - EyePhi.com

In the pentagram used to construct a golden spiral —and with which an ellipse can also be univocally determined in spherical geometry- we see that the reciprocal angles of the pentagon and the star are 108 and 72 degrees. On the other hand, Mathis himself comments, without relating it in any way to the optical equivalence, that in accelerators the relativistic mass of a proton usually finds a limit of 108 units that neither Relativity nor Quantum Mechanics explain, and he makes a derivation of the famous gamma factor that links it directly to G.

Of course, the Lorentz relativistic factor coincides with Weber's mechanics up to a certain limit of energy —although in the latter what increases is the internal energy instead of mass. There could be no more natural connection with the optical equivalence than that of light itself, and Mathis' theory establishes a series of equations and identities between light and charge, charge and mass, mass and gravity.

On the other hand, if we were to throw a stone into a well which perforated the Earth from side to side, and waited for it to return like a spring or a pendulum, it would take about 84 minutes, the same as an object in a close orbit around the planet. If we did the same thing with a particle of dust on an asteroid the size of an apple, but of the same density than our planet, the result would be exactly the same. This fact, which seems to assign an important role to density over mass and distance itself, pierce the appearance of the gravitational phenomenon, and should be as astonishing to us as Galileo's finding that objects fall at the same speed regardless of their weight; it also fits very well in the context of an spiral equal at all scales.

In any case the Lagrangian, the difference between kinetic and potential energy, has to play a fundamental role as a reference for the fine tuning of the different elements of the Solar System. In celestial mechanics, despite what is said, the integral has always led to the differential, and not the other way around. The law discovered by Newton does not shape the ellipse but rather tries to fit it.

So we have Newton's apple and the Golden Dragon of the Solar System Spiral. Will the dragon swallow the apple? The answer is that he doesn't need to swallow it, since it has been inside him from the start. Let us say it again: the gauge fields, characterized by the invariance of the Lagrangian under transformations, are equivalent to a non-trivial feedback between force and potential, which in turn is indistinguishable from the eternal "information problem", namely how the Moon knows where the Sun is and how it "knows" its mass to behave as it does. Why to ask about information at the microlevel of particles when the problem is in plain sight at the macrolevel in the first place?

Considering the adjustments of the Lagrangian in comparison with a system described exclusively by non-variable forces, the entire Solar System looks like a great spiral holonomy.

The Lagrangian can also hide virtual dissipation rates —virtual, of course, since we already know that the orbits are preserved.

[40]

In fact, what Lagrange did was to dilute D'Alembert's principle of virtual work by introducing generalized coordinates. But we are so used to separate the formalisms of thermodynamics from those of the supposedly more fundamental reversible systems that it is hard to see what this means. However, the most certain instinct tells us that everything reversible is an island surrounded by an ocean without forms. There is no motion without irreversibility; to pretend otherwise is just an illusion.

Mario J. Pinheiro wants to repair this divorce between convictions and formalisms by proposing a reformulation of mechanics alternative to the Lagrangian account, with a variational principle for rotating systems out of equilibrium and a mechanical-thermodynamic time in a set of two differential equations of first order. Here the equilibrium takes place between the minimum energy variation and the maximum entropy production.

This thermomechanics allows us to consistently describe systems with characteristics that are quite different from those of reversible systems, and which are particularly relevant to the case at hand: subsystems within a larger system can absorb the forces exerted on them, and instead of being enslaved there is room for interaction and self-regulation. There may be a component of topological torsion and conversion of linear or angular motion into angular motion. The angular momentum acts as a damper to dissipate the disturbances, "a well-known redressing mechanism in biomechanics and robotics " [15] .

To my knowledge, Pinheiro's proposal of an irreversible mechanics is the only one that gives a proper explanation of Newton's famous buck experiment and the whirlpool formed by its rotation, by the transport of angular momentum, as opposed to Newton's absolute interpretation or Leibniz's purely relational one, neither of which are really to the point. Suffice it to recall the elemental observation that in this experiment the appearance of the vortex requires both time and friction, and matter is transferred to the regions of highest pressure, a clear signature of the Second Law. What is extraordinary is that no one has insisted on this before

Pinheiro —something that can only be explained by the conventional roles assigned to the different branches of physics. Besides, it is clear that springs, whirls and spirals are the most suitable and efficient forms of damping.

It is perhaps appropriate to remember that the so-called "principle of maximum entropy" does not tend towards maximum disorder, as is often thought even among the physicists, but rather the opposite, and this is how Clausius originally understood it. This establishes a very broad but essential link with highly organized systems, at the top of which we usually place living beings [16]. On the other hand, it is enough to contemplate the spiral of the Solar System for a moment to understand that it only makes sense as an open, irreversible process in permanent production.

The concept of order that Boltzmann introduced is no less subjective than that of harmony, the main difference being that in statistical mechanics the micro-states, not the macro-states, have received a convenient quantification. Of course, this is another great rationalization: the irreversibility of phenomena or macro-states would be derived from the reversibility of microstates. But the mere postulation of stationary orbits in atoms —to pretend that there can be variable forces in isolated systems- is illegal both from the thermodynamic point of view and from the mere common sense.

The variational principle proposed by Pinheiro was first suggested by Landau and Lifshitz but has not been developed to date. This is inevitably reminiscent of the idea of damping wells in the Landau-Zener theory, which arise from adiabatic torque transfer when waves cross without destructive interference. Richard Merrick has directly related these wells or vortices to the golden spirals under conditions of resonance [17]. Many will say that one can not see how these conditions can be met in the Solar System, but, once again, the resonances of the classical theory of perturbations in Laplace's celestial mechanics are in no better situation, being nothing else than pure mathematical relations. If anything, it could be said that they are in a worse situation, since we are asked to believe that gravity can have a repulsive effect.

Although Pinheiro's thermomechanics involves something similar to this form of transfer, which evokes the parallel transport of the geometric phase, it also incorporates, and this is the key difference, a term for the thermodynamic free energy. A reversible system is a closed system, and there are no closed systems in the universe.

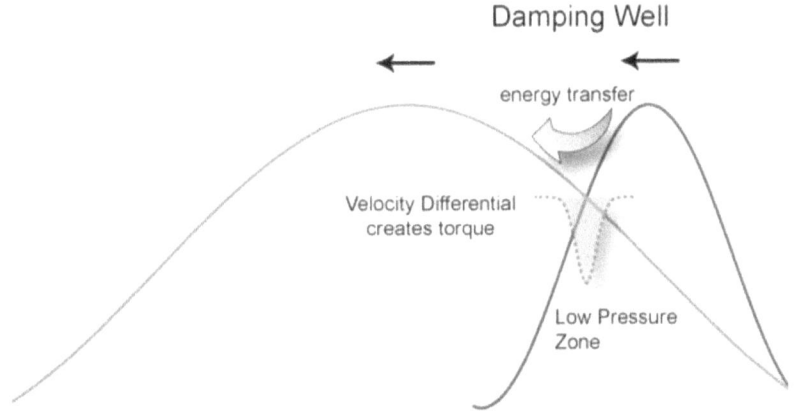

Richard Merrick: Harmonic formation helps explain why phi pervades the solar system

Merrick's own theory of harmonic interference would be elevated to a much higher level of generality simply by appreciating that the principle of maximum entropy production is not contrary to the generation of harmony but rather conducive to it.

The principle of maximum entropy can be transferred to quantum mechanics with hardly any more sacrifice than the idea of reversibility, as shown by the quantum thermodynamics developed by Beretta, Hatsopoulos and Gyftopoulos; the subject is of extraordinary importance but now it would take us too far [18].

*

Physicists are proud of the high degree of accuracy of some of their theories, which is quite understandable given the work invested carrying out their calculations, sometimes to ten and twelve decimal places. Few things would be more eloquent than such precision if it came naturally, without special assumptions or arbitrary ad hoc adjustments, but that is the case most of the time. Still the value of gravity on Earth cannot be measured to more than three decimal places, but astrophysicists pretend to calculate to ten or twelve places to the confines of the universe.

In the case of Lagrange and Laplace this is absolutely evident, and one day we will wonder how we were able to accept their methods without even blinking an eye. The truth is that these procedures were not digested overnight, but if they were finally accepted it was from the invincible desire to expand the power of calculus at any rate, reinforced by the idea, inherited from Newton and Leibniz, that Nature is a clockwork machine of virtually infinite precision. And for the means, what better than to serve the Ideal.

It has rightly been said that had Kepler had more precise data, he would not have advanced his theory of elliptical motion; and in fact, Cassini ovals, fourth-degree curves with a constant distance product, seem to reproduce the observable trajectories more closely, something one should attribute to the perturbations involved. These ovals also raise interesting and profound questions about the dynamical connection between ellipses and hyperbolas. Interestingly, Cassini ovals are used to model the geometry of the spontaneous negative curvature of red blood cells, in which the golden ratio has also been found [19].

As Mathis points out, the very first analyses of perturbations included, already since Newton and Clairaut, a factor $1/r^4$ with *a repulsive force*, which shows again to what extent the "auxiliary" elements of celestial mechanics are hiding something much more important [20].

To the eye of the naturalist, accustomed to the very variable precision of the descriptive sciences, the golden spiral of the Solar System would have to appear as the most splendid example of natural order; an order so magnificent that, unlike Laplace's, it can include catastrophes in its bosom without hardly blurring. This is a characteristic that we invariably attribute to living beings. Whether judged as a natural phenomenon or as an organism, taking everything into account, the spiral shows a precision, more than sufficient, excellent.

And what is the place of the *Taijitu*, our symbol of the Pole generating the yin and yang, in all this? Well, it goes without saying that the system we are talking about, along with its subsystems — planets and satellites —is an eminently polar process, with axes defining its evolution; and so it is the spiral holonomy that envelops them. As for the yin and yang, if we were to say that they can *also* be the kinetic and potential energy, we would be told that we are proposing too trivial a correspondence. But all the above should serve to see that this is not the case.

We know that in the orbits kinetic and potential energy do not even compensate, and when they should, as in the case of circular motion in Binet's equation, we do not even obtain a single force — at least a difference between the center of the circle and the center of the force is required. Looking for the simplest possible argument, the first thing that comes to mind is that the emergence of the golden section in the *Taijitu*, the freely rotating spherical vortex, contains a sort of analogical and a priori synthesis of 1) a law of areas applied to the two energies, 2) the focal geometry of the ellipses, and 3) a difference that is integrable and a shift in the plane of polarization that it is not. This third point overlaps the Lagrangian and a geometrical phase that in principle seem quite different.

Of course, we leave large loose ends here that such a simple diagram cannot translate. To begin with, just because an ellipse has two foci within does not mean that we have to look inside always for the origin of the forces that determine it, and this would lead us to the theory of perturbations. However, any environmental influence, also outer planets, should already be included in the geometric phase.

If we were to pass from celestial dynamics to light, we could reinterpret in terms of retarded potentials and their incidence on phase the data of ellipsometry or the "abstract monopole with a force of —1/2 at the center of the Poincaré sphere" to which Berry appeals in his generalization of the geometric phase. However, it should not be forgotten that light was already an essentially statistical process even since the days of Stokes and Verdet. The degree of polarization and entropy of a beam of light were always equivalent concepts, although we are still far from drawing all the consequences from this.

We assume the coincidence of the retarded potential and the geometrical phase, although there is not even a specific literature on the subject, nor is there agreement, otherwise, on the significance and status of the geometrical phase itself. There have been those who have seen it as an effect of the exchange of angular momentum, and in any case in classical mechanics the geometrical phase is shown by Hamilton-Jacobi's formulation of angle-action variables [21].

If harmony is totality, the so-called geometrical phase should have its part in the mathematics of harmony, since it is nothing but the expression of "global change without local change". We already noticed that the geometrical phase is inherent in fields involving conic sections, so its inclusion here is just elementary. However, the fact that it does not involve the known forces of interaction does not mean that we are dealing with mere "fictitious forces"; they are real forces that transport angular momentum and are essential in the effective configuration of the system.

Since this energy transport is an interference phenomenon, the potential energy of the Lagrangian must comprise the sum of all the interference from the adjacent systems, this being the missing "regulatory mechanism". It may be argued that in the course of the planets we do not observe the manifestation of interference that characterizes wave processes, even though we do not hesitate to resort to "resonances" to explain perturbations. Let us look at this a little more closely.

If until now we have chosen to see the geometrical phase, in classical mechanics the difference in the solid angle or Hannay

angle, as a relational property, the most appropriate way of understanding it would have to be within a purely relational mechanics
such as Weber's one. However, as Poincaré remarked, if we have
to multiply the velocity squared, *we no longer have a way of distin-
guishing between kinetic and potential energy, and even the latter
is no longer independent of the internal energy of the bodies* considered. Hence the postulation of an internal vibration by Noskov.
However, this inherent ambiguity does not prevent us from making
calculations as precise as with Maxwell's equations, in addition to
some other obvious advantages.

Remember the comparison of the stone that passes through
the Earth and the dust particle on that tiny asteroid, which return to
the same point in the same time. In a hypothetical medium of homogeneous density, this would suggest an overall dampening and
synchronizing effect at different spatial scales. But, without the
need for any hypothesis, what the geometrical phase implies is the
effective coupling of systems that evolve at different time scales,
for example, electrons and nuclei, or gravitational and atomic forces, or, within gravitational forces, the interactions between the different planets. This makes it particularly robust to noise or disturbances.

The ambiguity of relational mechanics need not be a weak-
ness, but could be revealing some limitations inherent in mechanics
and its calculus. Just when we want to take to its logical extreme
the ideal of converting physics into a pure kinematics, a science of
forces and motions, of mere extension, its inevitable dependence on
potentials and "non-local" factors is revealed, although we would
rather have to talk about definite global configurations.

What is essential in the apparently casual comparison between the *Taijitu* and the elliptical orbit is that the latter is also an
integral expression of the totality: not only of the internal forces
but also of the external forces that contribute to its form in real
time. If the compensation mechanism serves as an effective regulation it cannot affect only the potentials but equally the forces.

[47]

## 5. Questions of principle

The principles of Newtonian physics are based, as can be expected, in the circularity of its definitions of vector and scalar magnitudes like force and mass; the Lagrangian mechanics and the gauge fields, that "extend" it, demand a fixing to regulate the redundant degrees of freedom. Weber's law already allowed to appreciate in Kepler's problem the constituent elements of the gauge fields even if it dispensed completely with the very idea of fields —what changes here is the very status of the fundamental definitions, which are blurred. The retarded potentials allow to account for the essential aspects of modern physics, including the so-called relativistic effects.

The known and the unknown are easily confused with each other. On an immediate level, for us a force is, on the one hand, what induces motion, and on the other hand, what produces deformations in other bodies. But the "force" of gravity does not deform bodies when it forces them to move, and instead produce deformation when it does not force them —when it remains as a potential. Newton said that the whirling of the water buck spinning was due to centrifugal "fictitious forces" in absolute space, but Empedocles had shown two thousand years before that that same buck spinning above our heads counteracts the force of gravity.

How can one wield the axis of the Pole? The Pole, which balances the extremes of reality, is not something to be wielded. However the *Taijitu* invites us to look from its perspective. As for the spirit, Newton's three principles are tacitly summarized in the phrase "nothing moves unless it is moved" (by something else), that is, nothing moves without an external force —and neither relativity, quantum mechanics, nor modern cosmology have ever claimed otherwise. Everything is dead, except for the push given to it by something external. Now, all this prodigious development of modern science understood as mechanism is nothing but the unfolding of the consequences of the principle of inertia, and the ironic twist is that all the predictions of modern physics, and many others, can be made without the need of this principle at all.

The equivalence principle tells us that gravitational mass and inert mass are equal or indiscernible, and therefore the general theory of relativity states that there is no difference between the gravitational "force" and the fictitious forces. This is an attempt to walk in the direction of relational physics, but after terrible detours, and after arbitrating different versions —weakest, weak, middle-strong and strong- of such a principle, one ends up returning to the starting point, which is what it was all about.

The starting point is the principle of inertia, which no one wants to drop. The principle of inertia, often seen as purely redundant, bears the whole intentionality and disposition of the parts in physics. In other words, to do physics without the principle of inertia is equivalent to suspending its intention, which brings all operations back to the usual circular logic, with the help of the other two principles.

The principle of inertia, illustrated by the ball rolling for eternity in empty space, has been judged to be perfectly ideal. But it is not a perfect ideal, but a contradictory one: the motion of the ball must be related to axes of coordinates external to that system, and thus we have an isolated system that has the property of not being isolated. In reality there cannot be inertially isolated systems.

It is possible, and even necessary, as André Assis does, to propose a completely relational mechanics without using the concept of inertia, introducing instead the principle of dynamic equilibrium, so that "the sum of all forces of any nature acting on any body is always zero in all reference systems". This frees physics from the concepts of inertia, inert mass, absolute space, and the scholastic distinctions between frames of reference [22].

Something diametrically opposed to that implicit in the laws of mechanics also allows a description consistent with what we know. Thus, for example, Alejandro Torassa shows a dynamics valid for all observers in which "the motion of bodies is not determined by the forces acting on them, but the bodies themselves determine their motion" balancing the forces acting on them. "The natural state of a body in the absence of external forces is not only the state of rest or of uniform rectilinear motion, the natural state of movement of a body is any possible state of motion" [23].

If the sum of all forces is zero in any state, only differences and force ratios can be measured; introducing constants with dimensions here would be out of place. Another way of stating this principle would be to say that "the zero sum of all forces includes the observable motion," something that a certain mental inertia makes difficult to accept. Perhaps we can understand this better if we say that "observable motion balances the rest of the forces," that is, it balances out those that are not observable either. The balance of forces is not confused with their absence; but what we observe is motion and velocities, not forces.

There are of course other ways of representing this fundamental equilibrium without direct dependence on motion. For example, we can take from René Guenon the idea of an initially homogeneous medium, in which each compression at one point must correspond to an equal expansion at another point, so that their densities are reciprocal and their product equals always unity, even though the forces associated with them may be opposite, attractive or repulsive [24]. If in the originally homogeneous medium we imagine the correlative appearance of a fuller and an emptier portion, both could not simply arise without a torsion or helicity connecting them —and that torsion would be by itself a change in density. The characterization of equilibrium as a product is what we have considered here as reciprocity in the more intrinsic sense.

The cosmology of modern physics may argue that aspects such as general equilibrium are not matters of principle but of observation. What actually happens is that, if all that is observed is motion, and one starts from the principle of inertia, everything has to refer to causes external to what is observed —hence the hand of God to define the innate motion of the planets in Newton, or the notion of an event at the beginning of time that draws all energy from nothing. Thus, for example, and contrary to the endlessly publicized history, the first and most accurate predictions of microwave background radiation were not those of Gamow or other creationists, but those of physicists who assumed a universe in dynamic equilibrium [25].

This is the best example of the basic assumption being imposed on everything else, which instead tries to fit the assumption. It does not matter if this implies the most grandiose violation of the principle of energy conservation, as long as it is thrown out of bounds.

For modern physics, and not only physics, imbalance is the father of all things, and equilibrium is synonymous with death and disorganization. But the same observations and data have always allowed us to say that dynamic equilibrium is the father and mother of all things and that entropy does not lead to thermal death but to increased organization.

The true relevance of this dynamic equilibrium will be duly appreciated when mechanics and thermodynamics are united in a single discipline such as a thermomechanics of the type proposed by Pinheiro or other equivalents and more developed. And since his system of two equations is an alternative to the Lagrangian, it still fits better the Noskov equations, since the longitudinal vibrations of the bodies —which are congruent with Planck's formula- are equivalent to the input of free energy available in the medium.

[52]

Pinheiro's thermomechanics is conceived for open systems or systems out of equilibrium, and relational mechanics, even if the properties of the medium are not considered, as it lacks dimensional constants, depends implicitly on the environment.

# 6. Questions of interpretation - and of principle

According to Proclus, when Euclid wrote the *Elements* his primary objective was to elaborate a complete geometrical theory of the Platonic solids. Indeed, it has been said on several occasions that behind the name "Euclid" there could be a collective with a strong Pythagorean component. The existence of only five regular solids is possibly the best argument to think that we live in a three-dimensional world.

Another great champion of the concept of harmony in science was Kepler, the astronomer who introduced the ellipses into the history of physics —we only need to remember the title of his opus magnum, *Harmonices Mundi*. He discovered the convergence of Fibonacci's series, combined Pythagoras' theorem with the golden ratio in the triangle that bears his name, and theorized extensively about Platonic solids, which he even placed between the orbits of the planets, and which have been considered by a certain tradition as the arcana of the four elements plus the quintessence or Ether.

The latter would be represented by the pentagon and the pentagram, and by the dodecahedron dual of the icosahedron, corresponding to the element water.

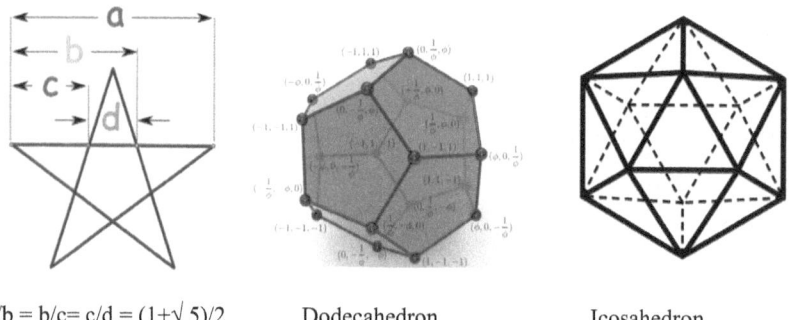

a/b = b/c= c/d = (1+√ 5)/2          Dodecahedron                    Icosahedron

The cascade of identical proportions at different scales immediately evokes the capacity to generate self-similar forms like the logarithmic spiral that displays its proportion indefinitely in a series of powers: $\varphi$, $\varphi^2$, $\varphi^3$, $\varphi^4$... If a recursion is that which allows something to be defined in its own terms, the pentagram shows us the simplest recursive process in a plane.

As late as 1884 Felix Klein, the reconciler of analytic and synthetic geometry, gave a group of *Lectures on the Icosahedron* in which he considered it to be the central object of the main branches of mathematics: "Every single geometrical object is connected in one way or another to the properties of the regular icosahedron". This is another of those "surprisingly well connected" objects and it would be interesting to trace the links with the two representations of the Pole from which we started. The Rogers-Ramanujan's continuous fraction, for example, plays a role for the icosahedron analogous to the exponential function for a regular polygon.

The orientation proposed by the influential Klein did not get as wide a reception as the famous Erlangen program; to properly deepen it required the dynamic guidance of nature, of applied physics and mathematics. Even current times are more propitious to recover this program, in spite of the fact that physics itself has become more abstract by leaps and bounds. The best way to revive Klein's second major program today might be with the use of a very down-to-earth version of geometric algebra.

The golden section emerges under the unequivocal seal of the fivefold symmetries, which were believed to be exclusive of living beings until the discovery of the quasi-crystals. That ordinary crystals, static structures by definition, exclude this type of symmetry seems to indicate long-range conditions of equilibrium. The optical properties of photonic quasi-crystals and other less periodic structures —often with a zero or double-zero refraction of permittivity and permeability- are being studied intensively in 1D, 2D, and 3D. As expected, phases with spiral holonomy have been found. As in graphene, although the Berry curvature has to be zero, the phase shift can be zero or $\pi$ [26].

It would be interesting to study all these new states from the point of view of thermomechanics, quantum thermodynamics and retarded potentials, the latter being able to offer a more convincing interpretation of, for example, the so-called relativistic effects of graphene; just as they allow to deal with singular points in a more logical way. On the border between periodic and random structures, the secrets of fivefold symmetry will not be opened without a careful, unbiased interpretation.

*

Except to confuse minds and things, the proscription of the Ether by the special relativity is anything but relevant. First of all, because what makes special relativity work is the Lorentz-Poincaré transformation, conceived expressly for the Ether. Second, because although special relativity, which is the general theory, dispenses with the Ether, general relativity, which is the special theory for gravity, demands it, even if it is in a rarefied theoretical way.

Weber's electrodynamics coincides closely with the Lorentz factor up to speeds of 0.85 $c$ without having to dispense with the third principle of mechanics. But in any case, and even if we stick to Maxwell's equations, which is the very crux of the matter, what we have is that there are not and cannot be electromagnetic waves moving in space with one component ortogonal to the other, but a statistical average of what occurs between space and matter. This is Mazilu's conclusion, but we only need to acknowledge the complete failure of any attempt to specify the geometric description of the field and the waves [27].

The strange thing, once again, is that this has not been seen clearly before. But it turns out that the idea of the Ether around 1900, at least the idea of Larmor, among others, was not only as a medium between particles of matter, but something that also penetrated those particles, which in fact were seen as condensations of the Ether —as now we can see the particles as condensations of the field. There was Ether outside in space and Ether inside matter — as in the electromagnetic waves, among which is light.

Ether is nothing but light itself, but it can also be other things than the light we see, and the whole electromagnetic spectrum. It's just that we cannot know anything without the help of light. Light is the mediator between a space that we cannot know directly, but gives us the metric, and a matter that is in the same situation, but is the subject to measurement.

And now that we know that we are in the midst of the Ether, like the bourgeois gentleman who discovered that he had always been speaking prose without knowing it, perhaps we can look at things more calmly. Only a consummately dualistic mentality that thinks in terms of "this or that" has been able to remain perplexed for so long on this question.

This idea of the Ether *in medias res* could not be very clear at the beginning of the 20th century, otherwise physicists would not have opened their arms to relativity as they did. If it was welcomed, leaving other reasons aside, it was because it seemed to end once and for all with an endless series of doubts and contradictions —or so they thought at the time, until insoluble "paradoxes" began to emerge one after another. One could think that by then Weber's law, which did not even need the existence of a medium because it did not consider waves either, had largely fallen into oblivion —though certainly not for researchers as exhaustive as Poincaré.

Of course, instead of Ether we can also use the word "field" now, as long as we do not understand it as the supplement of space that surrounds the particles, but as the fundamental entity from which they emerge.

In any case, special relativity proper practically does not come into contact with matter, and when it does through quantum electrodynamics, since Dirac, we are confronted with vacuum polarization and an even more crowded, strange and contradictory medium than in any of the previous avatars of the Ether.

Today, transformation optics and the anisotropies of metamaterials are used to "illustrate" black holes or to "design" —it is said- space-times in different flavors. And yet only the macroscopic parameters of Maxwell's old equations, such as permeability and permittivity, are being manipulated. And why should empty space have properties if it is really empty? But, again, what this is all about is statistical averages between space and matter. Only the prejudice created by relativity prevents us from seeing better these things.

It should be much more interesting to study the properties of the space-matter continuum accessible to our direct modulation than exotic aspects of hypothetical objects of a theory, the general relativity, which is not even unified with classical electromagnetism.

And if the ether of 1900 could prove to be inconvenient, what can we say about a theory that breaks with the continuity of the equations of classical mechanics, and that to alleviate it introduces infinite frames of reference? Certainly, it is not an economic solution, and it seems even worse if we think that with the principle of dynamic equilibrium we can dispense with inertia and the distinction of frames of reference —a move that is routinely used to rule out other theories and clear the field.

On the top of this, Maxwell's equations only are valid for extended portions of the field; special relativity is only valid for point-events, and in the field equations of general relativity point particles again become meaningless. Transformation optics takes advantage of this threefold incompatibility with a bypass that leaves special relativity in limbo, to link Maxwell with another incompatible theory. And yet, although this is not said, it is because of special relativity that quantum mechanics has been unable to work with extended particles. In contrast, starting from Weber mechanics there are no problems in working with both extended and point particles.

For those who still believe that the fundamental framework of classical mechanics must have four dimensions, it can be recalled that well into the 21st century, consistent gauge theories of gravity have been developed that satisfy the criterion formulated by Poincaré in 1902, namely to elaborate a relativistic theory in ordinary Euclidean space by modifying the laws of optics, instead of curving the space with respect to the geodesic lines described by light. Light is the mediator between space and matter; and if light can be deformed, which is obvious, there is no need to deform anything else.

Maxwell's equations are not even a general case, but a particular case, both of Weber and of Euler's equations of fluid mechanics. Within fluid mechanics, Maxwell sought the case for a static or motionless medium; if Maxwell's equations are not fundamental, the principle of relativity cannot be fundamental either [28]. The reciprocity of special relativity is purely abstract and kinematic, not mechanical, since it is not bound to centers of masses, and does not allow a distinction between internal forces that comply with the third law and external forces that do not need to comply with it. The principle of relativity, which asserts the impossibility of finding a privileged frame of reference, is valid if and only if there are no external forces to those considered within the system —but on the other hand, by neglecting the third principle, the internal forces are not defined mechanically either.

The so-called Poincaré stress that the French physicist introduced for the Lorentz force to comply with the third principle plays the same role in the relativistic context as Noskov's longitudinal vibrations for the Weber force. The fact that such stress was later considered irrelevant for special relativity shows conclusively its total divorce from mechanics.

[60]

Maxwell's equations, as Mazilu says, are a reaction to the partially or totally uncontrollable aspects of the Ether. In physical theories the quantities that matter are not those we can measure, but those we can control; but in this way we dispense with information that could be integrated into a broader theoretical framework.

Questions of interpretation inevitably bring us back to questions of principle; without changing the principles we are condemned to work for them.

The principle of relativity is contingent and therefore unnecessarily restrictive, depending also on arbitrary synchronization procedures. The equivalence principle of general relativity also does not put an end to the problems of reference frames, and, in combination with the principle of relativity, rather multiplies them.

The principle of dynamic equilibrium of relational mechanics radically simplifies this situation without creating unnecessary restrictions. Leaving generality aside, a principle should not be restrictive, but, first of all, necessary. On the other hand, the inability of the equivalence principle to get rid of the principle of inertia automatically subordinates it to the latter.

And it is no coincidence. If we know the same about inertia as we know about the Ether, it is rather because inertia itself inadvertently overlaps with the idea of the Ether and supplants it. Then, the Ether could only emerge without mystifications from a physics that completely dispenses with the idea of inertia, something that is perfectly feasible and compatible with all our experience.

Returning to the past, we see that Lorentz's non-dragging medium, Fresnel and Fizeau's partial dragging medium, and Stokes' total dragging medium are not contradictory and refer to clearly different cases. There are experiments, such as those of Miller, Hoek, Trouton and Noble, and many others, which can be carried out again under much better conditions and provide invaluable information from many points of view, provided that our theoretical framework allows us to contemplate them, which is not the case now [29]. In addition, these experiments are thousands of times less expensive, simpler and more informative than the current "confirmations" of special and general relativity.

There is also an inevitable complementarity between the constitutive aspects of electromagnetism in modern metamaterials, with their mixture of controllable and uncontrollable parameters in matter, and the measurement of uncontrollable parameters in a free environment out in space. But this complementarity cannot be appreciated without principles and a framework that can make them compatible, to begin with. On the other hand there is no need to say that between transformation optics and general relativity we have more of a flimsy parallelism than of real contact.

Another way of talking about a geometric phase is that it is a transformation or holonomy around a singularity. This singularity can be a vortex, which provides a natural connection with the entropy or attenuation of certain magnitudes, which obviously cannot reach infinite values.

An interesting case is the so-called transmutation of optical vortices, that is, the qualitative change of their most intrinsic feature, which is vorticity, and which has recently been carried out even in free space [30], also involving pentagonal symmetries. Vortices occur in the four states of matter —solid, liquid, gaseous and plasma, which are our version of the ancient four elements. Taking into account that their characteristic behavior can be described as a function of constitutive stress/strain relationships, a quantitative description of the transmutation of the states of matter is also feasible, which is widely different from the nuclear transmutation of the elements, without prejudice that also the nuclei can be described more or less classically with vortices like the skyrmions.

The so-called geometrical phase, this so universal phenomenon that it even manifests itself as vorticity on the surface of the water, applied to classical electromagnetism becomes, so to speak, "Maxwell's fifth equation", since it brings into play and encompasses the four known ones. The very name "geometric phase" seems clearly a euphemism, since it is not the geometricians who usually deal with it, but the physicists, and applied physicists for that matter. I also prefer to call it holonomy instead of anholonomy, since the latter refers to the fact that it cannot be integrated within the frame of a theory, while holonomy refers to a global aspect that can be recognized even by the naked eye.

Berry himself admits that the geometrical phase is a way of including the (uncontrollable) environmental factors that are not within the system defined by the theory [31]. In this sense, to get to Maxwell's "fifth equation" we don't need to add terms, but only refer to the "predynamics" of the less restricted or more general equations from which they originate —Weber's and Euler's in this instance. And the same applies to relativity.

No doubt the phenomenology of light is so vast that it never ceases to surprise us, but all this would have an incomparably greater transcendence if a parallel effort were dedicated to the uncontrollable but complementary aspects that are now outlawed or masked by the dominant theories. But in fact there is no part of physics that is not being contemplated today under an unnecessarily distorted optic.

*

There is also place in the theory of black holes for the golden ratio to emerge, just at the critical turning point when the temperature goes from rising to falling: $J^2/M^2 = (1+\sqrt{5})/2$, with M and J being the mass and angular momentum when the constants $c$ and G equals 1. The meaning and relevance of this is not clear, but it echoes the custom of this constant of appearing at critical points [32].

In Weber-type forces there is no room for theoretical objects such as black holes since the force decreases with the speed, and it would be interesting to see if transformation optics is able to find a laboratory replica for the evolution of these parameters. In relational mechanics the most that can be expected are different types of phase singularity, such as the optical vortices mentioned.

If there is any interest for us in the emergence of $\varphi$ in black holes, even if they are purely theoretical calculations, it is because of the direct association with angular momentum, entropy and thermodynamics. It shows us at least that the continuous proportion can also emerge in accordance with the principle of maximum entropy that we consider fundamental for understanding nature, quantum mechanics, or the thermomechanical formulation of classical mechanics. If it can emerge here, it can also do so in other types of singularities, such as the phase of vortices, in the optical model that de Broglie built for the light ray, or in holograms.

Black holes are extreme theoretical objects of maximum energy, but they have been reached through the "ordinary" physics of gravity, governed by action principles of minimum energy variation. Technically, this does not involve any contradiction, but makes us wonder about the very nature of the action principles, something that still worried a conservative physicist like Planck.

The action principle of Weber's law, or Noskov's extension of it, does not allow the existence of these extreme objects because, applying reciprocity in a strictly mechanical way to the centres of mass, force and speed get balanced. In the breaking down of the Lagrangian by Mathis into two forces, something similar happens. In Pinheiro's thermomechanics, in which there is a balance between minimum energy variation and maximum entropy, this does not seem to be possible either, provided there is free energy available.

The ordinary Lagrangian is the most void of causality, and Mathis' theory, which wants to dispense with energy and action principles in order to remain only with vectors and forces, would be obviously the «fullest» model; if it is viable is a different matter. The other two are in between. We know that with action principles univocal causes are impossible. One can choose the path one prefers and see how far it goes, but my position on this is that although a univocal determination of causes is not possible, we can have a statistical but certain sense of causality, related to the Second Law of Thermodynamics. Noskov's and Pinheiro's action principles seems compatible.

The third principle that defines what a closed system is, but the ironic twist is that this principle cannot be applied without the help of an environment with free energy contributing to close the balance. This happens even in Mathis' model, where free charge is recycled by matter. Thus, any reversible mechanics emerges as an island from an irreversible background, and it is this superposition of levels that gives us our intuition of causality.

The propagation of light is based on the homogeneity of space, but the masses on which gravity acts involve a non-homogeneous distribution. If we assume a primitive homogeneous medium, no type of force, including gravity, can alter that homogeneity except in a transitory way. The self-correction of forces, which is already implicit in Newton and the original Lagrangian, leads in that direction and seems the only conceivable way, if there is any, to cancel out the infinities arising in calculations. The entropic and thermodynamic treatment of gravity would also necessarily have to follow that direction.

The emergence of the continuous proportion and its series in plant growth, in pinecones or sunflowers, makes us think of vortices made up of discrete units, while at the same time it brings us back to the considerations about *the optimal, not maximum,* use of resources, matter and data collection that nature seems to display.

Yasuichi Horibe demonstrated that the Fibonacci binary trees were subject to the principle of maximum information entropy production [33], something that might be extended to thermodynamic entropy and, perhaps, to other branches such as optics, holography, or quantum thermodynamics. The question is whether these series can simply emerge from the principle of maximum entropy or are at some variable optimum point between minimum energy and maximum entropy, as suggested by Pinheiro's equations.

Since Pinheiro begins by testing its mechanics with some very elementary models, such as a sphere rolling on a concave surface, or the period of oscillation of an elementary pendulum, it would be of great interest to determine the simplest problem, within this mechanics, in which the continuous proportion appears with a critical or relevant role. With this we could take up again the Ariadne's thread of this ratio for action variables and optimization problems.

Planck was still concerned that the principles of action seem to imply a purpose. And the same was true of the Second Law of Thermodynamics for Clausius, although he was not at all bothered by that. It is plain to see that both types of processes, apparently so far apart, are effectively teleological, and this is not a coincidence since they are not even separated, as Pinheiro's thermomechanics shows. It seems that the simultaneous inclusion of two undeniable propensities of nature is more natural than their separate treatment.

In the West there has been a strong rejection of any teleological connotation because teleology has always been confused either with theology and the providential invisible hand or with the intentional hand of man. *Tertium non datur.* However, it is clear that here, for both mechanics and thermodynamics, we are talking about a tendency as unquestionable as it is spontaneous. Understanding this third position, which already existed before the false dilemma of mechanism, leads us to radically change our understanding of Nature.

# 7. Biological feedback —quantitative and qualitative models

In a recent paper we speculated on the presence of a geometric phase or phase memory in the bilateral nasal cycle, using a certain analogy between the mechanics of the circulatory system and a gauge field such as the electromagnetic one [34], and taking into account that Maxwell's equations are a particular case of the fluid equations. It is known that shortly after its discovery, the geometric phase was generalized well beyond the adiabatic or even the cyclic cases, and that today it is studied even in dissipative open systems and in various cases of animal locomotion. The analogy may be relevant despite the fact that the respiratory system obviously operates in a gaseous phase instead of a liquid one, while still being coupled to the blood circulation.

According to V. D. Tsvetkov, the ratio between systolic and diastolic time in humans and other mammals averages the same reciprocal values of the golden mean, and also the ratio of the maximum systolic pressure to the minimum diastolic pressure points to a relative value of 0.618/0.382 on average. Although these values may be arbitrarily approximate, we would have an excellent opportunity here to contrast them mechanically and see if there really is some kind of underlying optimization, since the systolic time already echoes the reflected vascular wave, and the same is true of the diastolic time.

On the other hand there is the Pulse Wave Velocity, which is a measure of arterial elasticity: both are derived from the second law of mechanics through the Moens-Korteweg equation. This wave velocity varies with pressure, as well as with the elasticity of the vessels, increasing with their stiffness. The return distance of the reflex wave and the time it takes increases with height, and a lower diastolic pressure, which indicates less resistance of the whole vascular system, reduces the magnitude of the reflected wave. Treatment of hypertension should focus, it is said, on decreasing the amplitude of the reflected wave, slowing it down, and increasing the distance between the aorta and the return points of this wave.

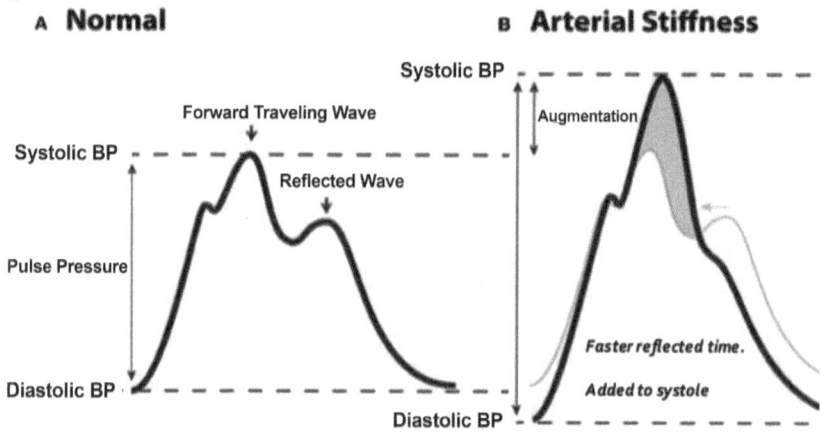

Now, we can try to apply here Noskov's retarded potentials with longitudinal waves, as he emphasized their universality and their place in the most elementary feedback; in fact, there is perhaps no better way of illustrating these waves and their correlation with certain proportions in a complete mechanical system than the circulatory system itself.

Since, to a large extent, it seems that we can consider the elasticity of the reflected wave as a retarded Weber-Noskov potential dependent on distance, force and phase velocity, and check whether this results in coupling or resonance conditions that incidentally tend to the values of the golden section. The myocardium is a self-exciting muscle, but the return of the reflex wave also contributes to this, so we have a fair example of a circuit with tension-pressure-deformation transformations that are fed back and that do not differ in essence from the gauge transformations of modern physics, in which there is also an implicit feedback mechanism.

This would be a perfect instance to explore these correlations in a sort of "closed loop" process, even if the system remains open through the breath, which is not contrary to our approach because for us every natural system is open by definition. It allows both numerical simulation and approximation by real physical models created with elastic tubes and coupled "pumps", so that it can be approached in the most tangible and direct way [35].

However, the idea that the heart is really a pump, being as it is a spiral muscular band, or that the motion of the blood, which generates vortices in the vessels and the heart, is due to pressure, when it is the pressure that is an effect of the former, should be thoroughly revised. In fact this is an excellent example of how we can give a strictly mechanical description while radically questioning not only the form but the very content of causality —the cause-effect relationship. The essential factor of the pressure created is not the heart, but the open component, in this case, the breath and the atmosphere. And although it is clear that these are very different cases, this is in line with our idea of gauge fields and natural processes in general.

*

The dynamics and biomechanics of the blood pulse can be derived from the applied force, but if we look within modern science for a suitable equivalent to Newton's three principles of mechanics in open systems such as biological organisms, we do not find it. To find something similar we have to look back to principles that are more "archaic" to us, and then look for a quantitative and mathematical translation.

Actually, the *triguna* of the Indian Samkya system —samkya means proportion- and its application to the human body as the *tri-dosha* in Ayurveda is the better match. The triguna, as it were, is a kind of system of coordinates for modalities of the material world in qualitative terms. The three basic qualities, *Tamas*, *Rajas* and *Satwa*, and their reactive forms in the body, *Kapha*, *Pitta* and *Vata* correspond very well to the mass or amount of inertia, the force or energy, and the dynamic equilibrium through motion (let us say: passivity, activity and balance). But it is evident that in this case we are talking about qualities and the systems are considered open from the start without need for further definition.

So, here is the law of conservation of momentum, not the third law of mechanics, what really should hold here, as a system like this implicitly admits a variable degree of interaction with the environment. In harmony with this, the Ayurveda considers that *Vata* is the guiding principle of the three since it has autonomy to move by itself in addition to moving the other two. *Vata* defines the sensitivity of the system in relation to the environment, its degree of permeability or lack thereof. In other words, the state of *Vata* indicates by itself to which extent the system is effectively open.

In the human body the most explicit and continuous form of interaction with the environment is the breath, and therefore it is just in the order of things that *Vata* governs this function most directly. Although the *doshas* are modes or qualities, in the pulse they find a faithful translation in terms of dynamic values and the continuum mechanics —provided we settle for modest degrees of precision, but surely enough to give us a qualitative idea of the dynamics and its basic patterns.

The other two modes are simply what moves and what is moved, but the articulation and coexistence of the three can be understood in very different ways: from a purely mechanical way to

[70]

a more specifically semiotic one. Here again, the indistinction or ambiguity between kinetic, potential and internal energy, which we have already noticed in relational mechanics, might be of some relevance.

The principle of inertia is a possibility, that of force a brute fact, the action-reaction —the same act seen from two sides- is a relationship of mediation or continuity. We can put them on the same plane or put them on different planes, which constitute an ascending or descending gradation, as in fact are the modalities of the Samkya system.

Actually, it should not be too difficult to find the common ground that the Indian and Chinese semiologies of the pulse have, beyond the differences of terminology and categories; and to move from this common ground to the quantitative, but extremely fluid, language of continuum mechanics. Thus we would have a method to pass from qualitative to quantitative aspects, and vice versa; and to find dynamic patterns that now pass unnoticed. There are several issues here. One is the extent to which these qualitative descriptions can be made consistent.

Another question is to what extent the representation of a qualitative scale can be made intuitive. Let us think, for example, of biofeedback signals, which can be effective under the representation of forces, potentials, and many other more indirect relationships. What is interesting is that these types of assisted feedback do not aim at control and manipulation, but at tuning in to the organizing principle of the dynamics.

From our perspective, as we have already said repeatedly, all physical systems, from galaxies to atoms, have feedback. But what are the physical limits of, let us say a human being, to tune in to other entities? The phase rhythmodynamics and its resonances, the time scale, the energy scale, strain-stress constitutive relations, the dependence on free energy? Or the capacity to align with the Pole that both systems have in common? Is there interference or is there rather a parallelism on the same background?

[71]

These are subjects for which science has not yet found even the minimum criteria, but which should help us to overcome the instrumental compulsion, the instrumentation syndrome that has guided human technology since the first tools, and which intensifies as the tools offers less resistance to the user.

One more question is whether this type of trimodal analysis, or even a bimodal one, has a recursive character, as the same feedback and the presence of the continuous proportion in the circulatory system suggest; and what type of recursion is involved.

*

The characterization of the dynamic equilibrium should always indicate the Pole of the evolution of a system, if it has one. In the case of the Solar System and the planets this is obvious —and notwithstanding, it is still far from receiving the attention it deserves. But it turns out that the bilateral nasal cycle is also telling us about an axis even in a process where polarity does not look very relevant, from the biomechanical point of view, such as the compression and release of a gas in our own organism. This should be of great interest to us, and it provides a thread through which many other things can be revealed.

In fact, the Earth's own climate or that of other planets, with its great complexity, is a more explicitly polar system than the respiratory regime of any mammal —and in this case the separating barrier would be the intertropical convergence zone. The point of interest here is that, if the analogy is sound, from a thermomechanical point of view the degree of separation that the barrier exerts, possibly associated with a topological torsion, could also be defining the degree of autonomy of the system with respect to the external conditions —let us call it the endogenous component. An endogenous view that would have to be duly complemented with appropriate sensors and observations of the so-called spatial time [36].

If we said before that the fact that an ellipse has in its interior two focuses does not mean that we only have to look inside it for the origin of the forces that determine its shape, the same is true for the disturbances that usually affect other organisms or systems, which does not prevent them from synthesizing in their behavior the product of external and internal factors, in the breath not less than in other balances that run in parallel.

# 8. Golden mean, statistics and probability

It has been said for some time that in today's science "correlation supersedes causation", and by correlation we obviously mean a statistical correlation. But even since Newton, physics has not been concerned that much with causation, nor could do it, so that more than a radical change we only have a steady increase in the complexity of the variables involved.

In the handling of statistical distributions and frequencies it makes little sense to talk about false or correct theories, but rather about models that fit the data better or worse, which gives this area much more freedom and flexibility with respect to assumptions. Physical theories may be unnecessarily restrictive, and conversely no statistical interpretation is truly compelling; but on the other hand, fundamental physics is increasingly saturated with probabilistic aspects, so the interaction between both disciplines continues to tighten.

Things become even more interesting if we introduce the possibility that the principle of maximum entropy production is present in the fundamental equations of both classical and quantum mechanics —not to mention if basic relations between this principle and the continuous proportion $\varphi$ were eventually discovered.

Possibly the reflected wave/retarded potential model we have outlined for the circulatory system gives us a good idea of a virtuous correlation/causation circle that meets the demands of mechanics but suspend —if not reverse- the sense in the cause-effect sequence. In the absence of a specific study in this area, we will now be content to mention some more circumstantial associations between our constant and the probability distributions.

The first association of φ with probability, combinatorics, binomial and hypergeometric distributions is already suggested by the presence of the Fibonacci series in the polar triangle already mentioned.

When we speak of probability in nature or in the social sciences, two distributions come first to mind: the almost ubiquitous bell-shaped normal or Gaussian distribution, and the power law distributions, also known as Zipf distributions, Pareto distributions, or zeta distribution for discrete cases.

Richard Merrick has spoken of a "harmonic interference function" resulting from harmonic damping, or in other words, the square of the first twelve frequencies of the harmonic series divided by the frequencies of the first twelve Fibonacci numbers. According to the author, this is a balance between spatial resonance and temporal damping.

In this way he arrives at what he calls a "symmetrical model of reflexive interference", formed from the harmonic mean between a circle and a spiral. Merrick insists on the transcendental importance for all life of its organization around an axis, which Vladimir Vernadsky had already considered to be *the* key problem in biology.

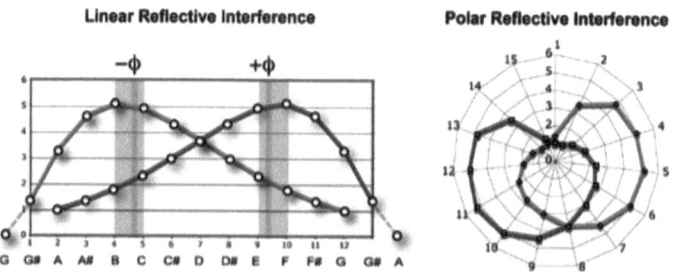

[76]

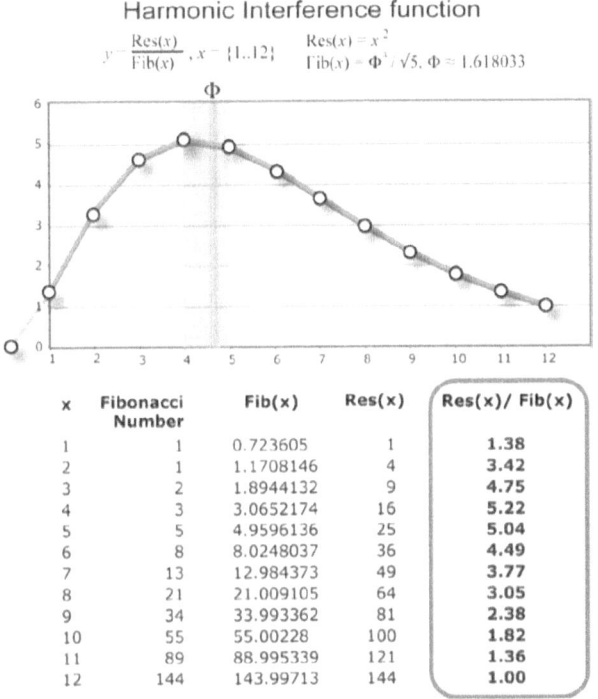

## Harmonic Interference function

$$y = \frac{Res(x)}{Fib(x)}, x = \{1..12\} \qquad Res(x) = x^2 \qquad Fib(x) = \Phi^x / \sqrt{5}, \Phi \approx 1.618033$$

| x | Fibonacci Number | Fib(x) | Res(x) | Res(x)/ Fib(x) |
|---|---|---|---|---|
| 1 | 1 | 0.723605 | 1 | 1.38 |
| 2 | 1 | 1.1708146 | 4 | 3.42 |
| 3 | 2 | 1.8944132 | 9 | 4.75 |
| 4 | 3 | 3.0652174 | 16 | 5.22 |
| 5 | 5 | 4.9596136 | 25 | 5.04 |
| 6 | 8 | 8.0248037 | 36 | 4.49 |
| 7 | 13 | 12.984373 | 49 | 3.77 |
| 8 | 21 | 21.009105 | 64 | 3.05 |
| 9 | 34 | 33.993362 | 81 | 2.38 |
| 10 | 55 | 55.00228 | 100 | 1.82 |
| 11 | 89 | 88.995339 | 121 | 1.36 |
| 12 | 144 | 143.99713 | 144 | 1.00 |

Richard Merrick, *Harmonically guided evolution*

Merrick's ideas about thresholds of maximum resonance and maximum damping can be put in line with Pinheiro's thermomechanical equations, and as we have indicated they would have a wider scope if they contemplated the principle of maximum entropy as conducive to organization rather than the opposite. Merrick elaborates also a sort of musical theory on the privileged proportion 5/6-10/12 at different levels, from the organization of the human torso to the arrangement of the double helix of DNA seen as the rotation of a dodecahedron around a bipolar axis.

\*

The powers laws and zeta distributions are equally important in nature and human events, and are present, among many other things, in fundamental laws of physics, the distribution of wealth among populations, the size of cities or the frequency of earthquakes. Ferrer i Cancho and Fernández note that "$\varphi$ is the value where the exponents of the probability distribution of a discrete magnitude and the value of the magnitude versus its rank coincide". It is not known at this time if this is a curiosity or if it will allow to deepen the knowledge of these distributions [37].

Zipf or zeta distributions are linked to hierarchical structures and catastrophic events, and also overlap with fractals in the space domain and with the so-called 1/f noise in the time domain. A. Z. Mekjian makes a broader study of the application of the Fibonacci-Lucas numbers to statistics that include hyperbolic power laws [38].

I. Tanackov et al. show the close relationship of the elementary exponential distribution with the value 2ln $\varphi$, which makes them think that the emergence of the continuous proportion in nature could be linked to a special case of Markov processes —a non-reversible case, we would advance. It is well known that exponential distributions have maximum entropy. It can be obtained an incomparably faster convergence to the value of the number $e$ with Lucas numbers, a generalization of Fibonacci numbers, than with Bernoulli's original expression, which is enough food for thought; we can also get with non-reversible walks a faster convergence than with the usual random walk [38].

Edward Soroko proposed a law of structural harmony for the stability of self-organized systems, based on the continuous proportion and its series considering entropy from the point of view of thermodynamic equilibrium [39]. Although here we give preference to entropy in systems far from equilibrium, his work is of great interest and can be a source of new ideas.

It would be desirable to further clarify the relationship of power laws to entropy. The use of the principle of maximum entropy seems to be particularly suitable for open systems out of balance and with a strong self-interaction. Researchers such as Matt Visser think that Jaynes' principle of maximum entropy allows a very direct and natural interpretation of powers laws [40].

Normally one looks for continuous power laws or discrete power laws, but in nature we can appreciate a middle ground between both as Mitchell Newberry observes with regard to the circulatory system. As usual, in such cases reverse engineering is imposed on the natural model. The continuous proportion and its series offer us an optimal recursive procedure to pass from continuous to discrete scales, and its appearance in this context could be natural [41].

The logarithmic average seems to be the most important component of these power laws, and we immediately associate the basis of the natural logarithms, the number $e$, with the exponential growth in which a certain variable increases without restrictions, something that in nature is only viable for very short periods of time. On the other hand, the golden mean seems to arise in a context of critical equilibrium between at least two variables. But this would lead us rather to logistic or S-curves, which are a modified form of the normal distribution and also a scaled compensation of a hyperbolic tangent function. On the other hand, exponential and power laws distributions look very different but sometimes can be directly connected, which is a subject on its own.

As already noticed, we can also connect the constants $e$ and $\Phi$ through the complex plane, as in the equality ($\Phi i = e^{\pm \pi i/3}$). Although entropy has always been measured with algebras of real numbers, G. Rotundo and M. Ausloos have shown that here too the use of complex values can be justified, allowing to treat not only a "basic" free energy but also "corrections due to some underlying scale structure" [42]. The use of asymmetric correlation matrices could also be linked with the golden matrices generalized by Stakhov and applied to genetic code information by Sergey Pethoukov [43].

In the mechanical-statistical context the maximum entropy is only an extreme referred to the thermodynamic limit and to Poincaré's immeasurable scales of recurrence; but in many relevant cases in nature, and evidently in the thermomechanical context, it is necessary to consider a non-maximum equilibrium entropy, which may be defined by the coarse grain of the system. Pérez-Cárdenas et al. show a non-maximum coarse-grained entropy linked to a power law, the entropy being so much lower when finer is the grain of the system [44]. This graininess can be linked to the constants of proportionality in the equations of mechanics, such as the same Planck's constant.

*

Probability is a predictive concept and statistics a descriptive, interpretative one, and both should be balanced if we do not want human beings to be increasingly governed by concepts they do not understand at all.

Just to give an example, the mathematical apparatus known as the renormalization group applied to particle and statistical physics is particularly relevant in deep learning, to the point that some experts claim both are the same thing. But it goes without saying that this group historically emerged to deal with the effects of the Lagrangian self-interaction in the electromagnetic field, a central theme of this article.

For prediction, the effects of self-interaction are mostly "pathological", since they complicate calculations and often lead to infinity —although in fact we should put the blame for this in the inability to work with extended particles of special relativity, rather than in self-interaction. But for the description and interpretation the problem is the opposite, it is about recovering the continuity of a natural feedback broken by layers and more layers of mathematical tricks. The conclusion could not be clearer: the search for predictions, and the "artificial intelligence" thus conceived, has grown exponentially at the expense of ignoring natural intelligence —the intrinsic capacity for self-compensation in nature.

If we want to somehow reverse the fact that man is increasingly governed by numbers that he does not understand —and even the experts are bound to trust in programs whose outputs are way beyond their understanding- it is necessary to work at least as hard in a regressive or retrodictive direction. If the gods destroy men by making them blind, they make them blind by means of predictions.

*

As Merrick points out, for the current theory of evolution, if life were to disappear from this planet or have to start all over again, the long-term results would be completely different, and if a rational species were to emerge it would be totally different from our own. That is what random evolution means. In a harmonically guided evolution conditioned by resonance and interference as Merrick suggests, the results would be fairly the same, except for the uncertain incidence that the great cosmic cycles beyond our reach might have.

There is not something like pure chance, there is nothing purely random; no matter how little organized an entity is, be it a particle or an atom, it cannot fail to filter out the environmental "random" influences according to its own intrinsic structure. And the first sign of organization is the appearance of an axis of symmetry, which in particles is defined by axes of rotation.

The dominant theory of evolution, like cosmology, has emerged to fill the great gap between abstract and reversible, and therefore timeless, physical laws and the ordinary world showing an irreversible time, perceptible forms and sequences of events. Today's whole cosmology is really based on an unnecessary and contradictory assumption, the principle of inertia. The biological theory of evolution is based on a false one, that life is only governed by chance.

The present "synthetic theory" of evolution has only come into existence because of the separation of disciplines, and more specifically, due to the segregation of thermodynamics from fundamental physics despite the fact that there is nothing more fundamental than the Second Law. It is not by chance that thermodynamics emerged simultaneously with the theory of evolution: the first

[81]

one begins with Mayer, who elaborated on work and physiology considerations, and the second one with Wallace and Darwin starting, according to the candid admission of the latter in the first pages of his main work, from Malthus' assumptions of resources and competition, which in turn go back to Hobbes —one is a theory of work and the other of the global ecosystem understood as a capital market. The accumulated capital in this ecosystem is, of course, the biological inheritance.

Merrick's harmonic evolution, due to the collective interference of waves-particles, is an updating of an idea as old as music; and it is also a timeless, purpose-free vision of the events of the world. But to reach the desired depth in time, it must be linked to the other *two clearly teleological, but spontaneous domains,* of mechanics and thermodynamics, which we call thermomechanics for short.

It would be enough to unite these three elements for the present theory of evolution to start becoming irrelevant; and not to mention that human and technological evolution is decidedly Lamarckian beyond speculation. Even DNA molecules are organized in the most obvious way along an axis. And as for information theory, one only has to remember that it has come out of a peculiar interpretation of thermodynamics, and that it is impossible to do automatic computations without components with a turning axis. Whatever the degree of chance, the Pole rules and defines its sense and meaning.

However, in order to better understand the action of the Pole and the spontaneous reaction involved in mechanics it would be good to rediscover the meaning of polarity.

## 9. Questions of program —and of principle, again: calculus, dimensional analysis and chronometrology

In physics and mathematics, as in all areas of life, we have principles, means and ends. The principles are the starting points, the means, from a practical-theoretical point of view, are the different branches of calculus, and the interpretations the ends. These last ones, far from being a philosophical luxury, are the ones that determine the whole contour of representations and applications of a theory.

As for principles, we have already commented, if we want to see more closely where and how the continuous proportion emerges, we should observe as much as possible the ideas of continuity, homogeneity and reciprocity. And this includes the consideration that all systems are open, since if they are not open they cannot comply with the third principle in a way that is worthy of being considered "mechanical". This is the main difference between Nature and manmade machines.

These three-four principles, are nonspecifically included in the principle of dynamic equilibrium, which is the way to dispense with the principle of inertia, and incidentally, the principle of relativity. If we talk about continuity, this does not mean that we are stating that the physical world must necessarily be continuous, but that what seems to be a natural continuity should not be broken without need.

Actually, the principles also determine the scope of our interpretations, although they do not specify them.

As for calculus, which in the form of predictions has become for modern physics the almost exclusive purpose, it is always a matter of justifying how to achieve results known in advance, so reverse engineering and heuristics always won the day over considerations of logic or consistency. Of course there is a laborious foundation of calculus by Bolzano, Cauchy and Weierstrass, but it is more concerned with saving the results than with making them more intelligible.

On this point we cannot agree more with Mathis, who stands alone in a battle to redefine these foundations. What Mathis proposes can be traced back to umbral calculus and the calculus of finite differences, but these are considered as sub-domains of the standard calculus and ultimately have not brought a better understanding to the field.

An instantaneous speed is still an impossible that reason rejects, and besides there is no such thing on a graph. If there are physical theories, such as special relativity, that unnecessarily break with the continuity of classical equations, here we have the opposite case, but with another equally disruptive effect: a false notion of continuity, or pseudo-continuity, is created that is not justified by anything. Modern calculus has created for us an illusion of dominion over infinity and motion by subtracting at least one dimension from the physical space, not honoring that term "analysis" which is so proud of. And this, naturally, should have consequences in all branches of mathematical physics [45].

Mathis' arguments are absolutely elementary and irreducible; these are also questions of principle, but not only of principle as the problems of calculus are eminently technical. The original calculus was designed to calculate areas under curves and tangents to those curves. It is evident that the curves of a graph cannot be confused with the real trajectories of objects and that in them there are no points or moments; then all the generalizations of this methods contain this dimensional conflation.

Finite calculus is also closely related to the problem of particles with extension, without which it is nearly impossible to move from the ideal abstraction of physical laws to the apparent forms of nature.

Mathis himself is the first to admit, for example in his analysis of the exponential function, that there is still a great deal to do for a new foundation of calculus, but this should be good news. At any rate the procedure is clear: the derivative is not to be found in a differential that approaches zero, nor in limit values either, but in a sub-differential that is constant and can only be 1, a unit interval. A differential can only be an interval, never a point or subtraction of points, and it is to the interval that the very definition of limit owes its range of validity. In physical problems this unit interval must correspond to an elapsed time and a distance traveled.

Trying to see beyond Mathis' efforts, it could be said that, if curves are defined by exponents, any variation in a function should be able to be expressed in the form of a dynamic equilibrium whose product is unity; and in any case by a dynamic equilibrium based on a constant unit value, which is the interval. If classical mechanics and calculus grew up side by side as almost indistinguishable twins, even more so should be in a relational mechanics where inertia always dissolves in motion.

The heuristic part of modern calculus is still based on averaging or error compensation; while the foundation is rationalized in terms of limit, but works due to the underlying unit interval. The parallel between the bar of a scale and a tangent is obvious; what is not seen is precisely what should be compensated. Mathis method does not work with averages; standard calculus does. Mathis has found the beam of the scale, now it comes down to set the plates and fine tune the weights. We will come back to this later.

The disputes that still exist from time to time, even among great mathematicians, regarding standard and non-standard calculus, or the various ways of dealing with infinitesimals, at least reveal that different paths are possible, but for most of us they are remote discussions far removed from the most basic questions that should be analyzed first.

We mentioned earlier Tanackov's formula to calculate the constant $e$ way faster than the classic "direct method"; but the fact is that amateur mathematicians like the already mentioned Harlan Brothers have found, just twenty years ago, *many* different closed expressions that calculate it faster, besides being more compact. The mathematical community may treat it as a curiosity, but if this happens with the most basic rudiments of elementary calculus, what cannot happen in the dense jungle of higher order functions.

A somewhat comparable case would be that of symbolic calculus or computer algebra, which already 50 years ago found that many classical algorithms, including much of linear algebra, were terribly inefficient. However, as far as we can see, none of this has affected calculus proper.

"Tricks" like those of Brothers play in the ground of heuristics, although it must be recognized that neither Newton, nor Euler, nor any other great name of calculus knew them; but even if they are heuristics they cannot fail to point in the right direction, since simplicity use to be indicative of truth. However with Mathis we are talking not only about the very foundations, which none of the revisions of symbolic calculus has dared to touch, but even about the validity of the results, which cross the red line of what mathematicians want to consider. At the end of the chapter we will see whether this can be justified.

In fact, to pretend that a differential tends to zero is equivalent to say that everything is permitted in order to get the desired result; it is the ideal condition of versatility for whatever heuristics and adhocracy. The fundamental requirement of simplified or unitary calculus —plain differential calculus, indeed- may at first seem like putting spanners in the works already in full gear, but nonetheless it is truthful. No amount of ingenuity can replace rectitude in the search for truth.

[86]

Mathis' attempt is not quixotic; there is here much more than meets the eye. There are reversible standards and standards irreversible in practice, such as the current inefficient distribution of the letters of the alphabet on the keyboard, which seems impossible to change even though nobody uses the old typewriters anymore. We do not know if modern calculus will be another example of a standard impossible to reverse, but what is at stake here is far beyond questions of convenience, and it blocks a better understanding of an infinite number of issues; overcoming it is an indispensable condition for the qualitative transformation of knowledge and the ideas we have —or cannot have- of space, time, change and motion.

*

Today's theoretical physicists, forced into a highly creative manipulation of the equations, tend to dismiss dimensional analysis as little more than pettifogging; surely this attitude is due to the fact that they have to think that any revision of the foundations is out of question, and one can only look ahead.

In fact, dimensional analysis is more inconvenient than anything else, since it is by no means irrelevant: it can prove with just a few lines that the charge is equivalent to mass, that Heisenberg's uncertainty relations are conditional and unfounded, or that Planck's constant should only be applied to electromagnetism, instead of being generalized to the entire universe. And since modern theoretical physics is in the business of generalizing its conquests to everything imaginable, any contradiction or restriction to its expansion by the only place it is allowed to expand must be met with notorious hostility.

It is actually easy to see that dimensional analysis would have to be a major source of truth if it were allowed to play its part, since modern physics is a tower of Babel of highly heterogeneous units that are the reflection of contortions made in the name of algebraic simplicity or elegance. Maxwell's equations, compared to the Weber force that preceded him, are the most eloquent example of this.

[87]

Dimensional analysis is also of interest when we delve into the relationship between intensive and extensive quantities. The disconnection between the mathematical constants $e$ and $\varphi$ may also be associated with this broad issue. In entropy, for example, we use logarithms to convert intensive properties such as pressure and temperature into extensive properties, converting for convenience multiplicative relations into more manageable additive relations. That convenience becomes a necessity only for those aspects that are already extensive, such as an expansion.

Ilya Prigogine showed that any type of energy is made up of an intensive and an extensive variable whose product gives us an amount; an expansion, for example, is given by the PxV product of pressure (intensive) by volume (extensive). The same can be applied to changes in mass/density with velocity and volume, and so on.

The unstoppable proliferation of measurements in all areas of expertise already makes simplification increasingly necessary. But, beyond that, there is an urgent need to reduce the heterogeneity of physical magnitudes if intuition is to win the battle against the complexity with which we are accomplices.

All this is also closely related to finite calculus and the equally finitist algorithmic measurement theory developed by A. Stakhov. The classical mathematical measure theory is based on Cantor's set theory and as we know it is neither constructive nor connected with practical problems, let alone the hard problems of the modern physical measurement theory. However, the theory developed by Stakhov is constructive and naturally incorporates an optimization criterion.

To appreciate the scope of the algorithmic measurement theory in our present quantitative Babel we must understand that it takes us back to the Babylonian origins of the positional numbering system, filling an important gap in the current theory of numbers. This theory is isomorphic with a new number system and a new general theory of biological population. The number system, created by George Bergman in 1957 and generalized by Stakhov, is based on the powers of φ. If for Pythagoras "all is number", for this system "all number is continuous proportion".

The algorithmic measure theory also raises the question of equilibrium, since its starting point is the so-called Bachet-Mendeleyev problem, which curiously also appears for the first time in Western literature with Fibonacci's *Liber Abacci* of 1202. The modern version of the problem is to find the optimal system of standard weights for a balance that has a response time or sensitivity.

According to Stakhov, the key point of the weight problem is the deep connection between measurement algorithms and positional numbering methods. My impression however is that it still supports a deeper connection between dynamic equilibrium, calculus and what it takes to adjust a function; the weights for the plates needed by the beam of the scale identified by Mathis. Maybe there is no need to use powers of φ or to change the number system, but very useful ideas might be developed about the simplest algorithms.

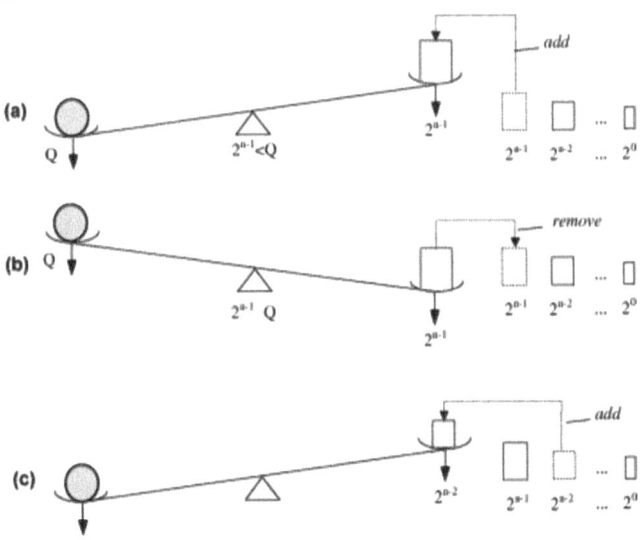

Alexey Stakhov: algorithmic measurement theory

[89]

Of course, the algorithmic theory of complexity tells us that we cannot prove an algorithm is the simplest for a task, but that does not mean that we do not look for it continually, regardless of any demonstration. Efficiency and formal demonstration are largely unrelated.

Human beings inevitably tend to optimize what they measure most; however, we do not have a theory that harmonizes the needs of metrology with those of mathematics, physics and the descriptive sciences, whether social or natural. Today there are many different measure theories, and each discipline look for what is best for it. However, all metrics are defined by a function, and functions are defined by calculus or analysis, which does not want to have anything to do with the practical problems of measurement and pretends to be as pure as arithmetic even though it is far from it.

This may seem a somewhat absurd situation and in fact it is, but it also places a hypothetical measure theory that is in direct contact with the practical aspects, the foundations of calculus and arithmetic in a strategic situation above the drift and inertia of the specialities.

Calculus or analysis is not pure math, and it is too much to pretend that it is. On the one hand, and as far as physics is concerned, it involves at least a direct connection with questions of measurement that should be more explicit in the same math; on the other hand, the highly heuristic nature of its most basic procedures speaks for itself. If arithmetic and geometry have large gaps, being incomparably clearer, it would be absurd to pretend that calculus cannot have gaps much greater.

*

On the other hand, physics will never cease to have both statistical and discrete components —bodies, particles, waves, collisions, acts of measurement, etc- besides continuous ones, which makes a relational-statistical analysis advisable.

An example of relational statistical analysis is the one proposed by V.V. Aristov. Aristov introduces a constructive and discrete model of time as motion using the idea of synchronization and physical clock that Poincaré already introduced precisely with the problem of the electron. Here each moment of time is a purely spatial picture. But it is not only a matter of converting time into space, but also of understanding the origin of the mathematical form of the physical laws: "The ordinary physical equations are consequences of the mathematical axioms, 'projected' into physical reality by means of the fundamental instruments. One can assume that it is possible to build different clocks with a different structure, and in this case we would have different equations for the description of motion".

Aristov himself has provided clock models based on non-periodic, theoretically random processes, that are also of great interest. A clock based on a non-periodic process could be, for example, a piston engine in a cylinder; and this could also include thermodynamic processes.

It should also be noted that cyclical processes, despite their periodicity, mask additional or environmental influences, as we have seen with the geometric phase. To this, a deductive filter of unnecessarily restrictive principles is added, as we have already seen in the case of relativity. And as if all this were not enough, we have the fact, hardly recognized, that many processes considered purely random or "spontaneous", such as radioactive decay, show discrete states during fluctuations in macroscopic processes, as has been extensively shown by S. Shnoll and his school for more than half a century.

Indeed, all kind of processes, from radioactive decay to enzymatic and biological reactions, through random number generators, show recurrent periods of, 24 hours, 27 and 365 days, which obviously correspond to astronomical cyclic factors.

We know that this regularity is filtered and routinely discounted as "non-significant" or irrelevant, in an example of how well researchers are trained to select data, but, beyond this, the question of whether such reactions are spontaneous or forced remains. An answer may be advanced: one would call them spontaneous even if

[91]

a causal link could be demonstrated, since bodies contribute with their own momentum.

The statistical performance of multilevel neural networks — ultimately a brute force strategy- is increasingly hampered by the highly heterogeneous nature of the data and units with which they are fed, even though dynamic processes are obviously independent of the units. In the long run, the pureness of principles and criteria is irreplaceable, and the shortcuts that theories have sought for prediction accumulates a huge deal of deadweight. And again, it is of little use what conclusions machines can reach when we are already incapable of seeing through the simplest assumptions.

The performance of a relational network is also cumulative, but in exactly the opposite sense; perhaps it should be said, rather, that it grows in a constructive and modular way. Its advantages, such as those of relational physics —and information networks in general - are not obvious at first sight but increase with the number of connections. The best way to prove this is by extending the network of relational connections. And indeed, it is about collective work and collective intelligence.

With arbitrary cuts to relational homogeneity, destructive interference and irrelevant redundancy increase; conversely, the greater the relational density, the greater the constructive interference. I don't think this requires demonstration: Totally homogeneous relations allow higher order degrees of inclusion without obstruction, just as equations made of heterogeneous elements can include equations within equations as opaque elements or knots still to unravel [46].

*

Let us go back to the calculus, but from a different angle. Mathis' differential calculus does not always get the same results as the standard one, which would seem sufficient to rule it out. Since the principle is unquestionable, errors, if possible, might be due to an incorrect application of the principle, leaving the criteria still to be clarified. On the other hand, that there is a "dimensional reduction" of the curves in standard calculus is a fact, which howe-ver is not widely recognized because after all now the graphs are supposed to be secondary and even dispensable.

Are they really? Without graphs and curves calculus would never have been born, so that is enough. David Hestenes, the great advocator of geometric algebra and geometric calculus, says that geometry without algebra is dumb, and algebra without geometry is blind. We should add that not only algebra, but calculus too, and to a greater extent than we think, provided we understand that there is more to "geometry" than what the graphs usually tell us. We can now look at another type of graph, this time of vortices, due to P. A. Venis [47].

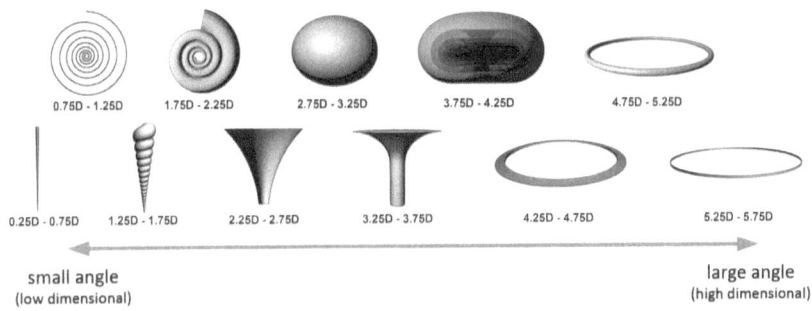

Peter Alexander Venis

In the transformation sequence Venis makes an estimate of its dimensionality that at first may seem arbitrary, although it is based on something as "obvious" as the transition from point to line, line to plane and plane to volume. The fractional dimensions seems at first sight striking, until we recognize it is just an estimate of the continuity within the order of the sequence, which could not be more natural.

Although Venis does not look for a proof, his transformation sequence is self-evident and more compelling than a theorem. One need only a minute of real attention to understand its evolution. It is a general key to morphology, regardless of the physical interpre-tation we want to give it.

For Venis the appearance of a vortex in the physical plane is a phenomenon of projection of a wave from a single one field where the dimensions exist as a compact whole without parts: a different way to express the primitive homogeneous medium of reference for dynamic equilibrium. It is clear that in a completely homogeneous medium we cannot characterize it as either full or void, and that we could say that it has either an infinite number of dimensions or no dimensions at all.

Thus, both the ordinary dimensions, as well as the fractional ones, and even the negative dimensions are a phenomenon of projection, of projective geometry. The physical nature is real since it participates in this one field or homogeneous medium, and it is a projected illusion to the extent that we conceive it as an independent part or an infinity of separate parts.

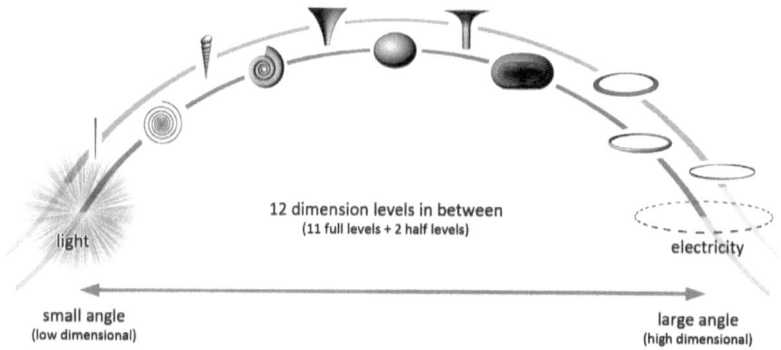

Peter Alexander Venis

Negative dimensions are due to a projection angle lower than 0 degrees, and lead to toroidal evolution beyond the bulb in equilibrium in three dimensions —that is, dimensions greater than the ordinary three. So they form a complementary projective counter-space to the ordinary space of matter, which with respect to unity is not less a projection than the first. Light and electricity are at opposite ends of manifestation, of evolution and involution in matter: light is the fiat, and electricity the extinction. Much could be elaborated on this but we will leave something for later.

Arbitrary cuts in the sequence leave fractional dimensions exposed, coinciding with the shapes we can appreciate. Since Mathis himself attributes the differences in results between his calculus and the standard calculus to the fact that the latter eliminates at least one dimension, and in the sequence of transformations we have a whole series of intermediate dimensions for basic functions, this would be an excellent workbench to compare both.

Michael Howell consider that fractal analysis avoids the usual dimensional reduction, and translates the exponential curve into "a fractal form of variable acceleration" [48]. It is worth noting that for Mathis the standard calculus has errors even in the elementary exponential function; the analysis of the dimensional evolution of vortices gives us a wide spectrum of cases to settle differences. I am thinking about fractional derivatives and differentiable curves, rather than fractals as non-differentiable curves. It would be interesting to see how the constant differential works with fractional derivatives.

The history of fractional calculus, which has gained great momentum in the 21st century, goes back to Leibniz and Euler and is one of the rare cases where both mathematicians and physicists ask for an interpretation. Although its use has extended to intermediate domains in exponential, wave-diffusion, and many other types of processes, fractional dynamics presents a non-local history dependence that deviates from the usual case, though there is also local fractional calculus. To try to reconcile this divergence Igor Podlubny proposed a distinction between inhomogeneous cosmic time and homogeneous individual time [49].

Podlubny admits that the geometrization of time and its homogenization are primarily due to calculus itself, as the intervals of space can be compared simultaneously, but the intervals of time cannot, and we can only measure them sequentially. What may be surprising is that this author attributes non-homogeneity to cosmic time, rather than to individual time, since in reality mechanics and calculus develop in unison under the principle of global synchronization and simultaneity of action and reaction. In this respect relativity is not different from Newtonian mechanics. According to Podlubny, individual time would be an idealization of the time created by mechanics, which is to put it upside down: in any case the idealization is the global time.

On the one hand, fractional calculus is seen as a direct aid for the study of all kinds of "anomalous processes"; on the other hand, fractional calculus itself is a generalization of standard calculus that includes ordinary calculus and therefore also allows to deal with all modern physics without exception. This makes us wonder if, more than dealing with anomalous processes, it is ordinary calculus what enforces a normalization, which affects all quantities it computes, time among them.

Venis also speaks of non-homogeneous time and temporal branches, though his reasoning remains undecided between the logic of the sequence, which represents an individualized flow of time, and the logic of relativity. However, it is the sequential logic that should define time in general and individual or local time in particular —not the logic of simultaneity of the global synchronizer. We shall return to this soon.

## 10. From monopoles to polarity

Polarity was always an essential component of natural philosophy and even of plain thought, but the advent of the theory of electric charge replaced a living idea with a mere convention.

Regarding the universal spherical vortex, we mentioned earlier Dirac's monopole hypothesis. Dirac conjectured the existence of a magnetic monopole just for the sake of symmetry: if there are electric monopoles, why can not exist magnetic monopoles too?

Mazilu, following E. Katz, suggest quietly that there is no need to complete this symmetry, since we already have a higher order symmetry: the magnetic poles appear separated by portions of matter, and the electric poles only appear separated by portions of space. This is in full accordance with the interpretation of electromagnetic waves as a statistical average of what occurs between space and matter.

And this puts the finger on the point everybody tries to avoid: it is said that the current is the effect that occurs between charges, but actually it is the charge what is defined by the current. Elementary charge is a postulated entity, not something that follows from definitions. Mathis can rightly say that the idea of elementary charge can be completely dispensed with and replaced by mass, which is justified by dimensional analysis and greatly simplifies the picture —freeing us, among other things, from "vacuum constants" such as permittivity and permeability that are completely inconsistent with the word "vacuum" [50].

In this light, there are no magnetic monopoles, because there are no electric monopoles nor dipoles to begin with. The only things that would exist are gradients of a neutral charge, photons producing attractive or repulsive effects according to the relative densities and the screening exerted by other particles. And by the way, it is this purely relative and changing sense of shadow and light what characterized the original notion of yin and yang and polarity in all cultures until the great invention of elementary charge.

So it can well be said that electricity killed polarity, a death that will not last long since polarity is a much larger and more interesting concept. It is truly liberating for the imagination and our way of conceiving Nature to dispense with the idea of bottled charges everywhere.

Yes, it is more interesting even for such a theoretical subject as the monopole. Theoretical physicists have even imagined global cosmological monopoles. But it is enough to imagine a universal spherical vortex like the one already mentioned, without any kind of charge, but with self-interaction and a double motion for the rotations associated with magnetism and the attractions and repulsions associated with charges to arise. The same reversals of the field in Weber's electrodynamics already invited to think that the charge is a theoretical construct.

We should come to see electromagnetic attraction and repulsion as totally independent of charge, and conversely, the unique field that includes gravity, as capable of both attraction and repulsion. This is the condition, not to unify, but to approach the effective unity that we presuppose in Nature.

*

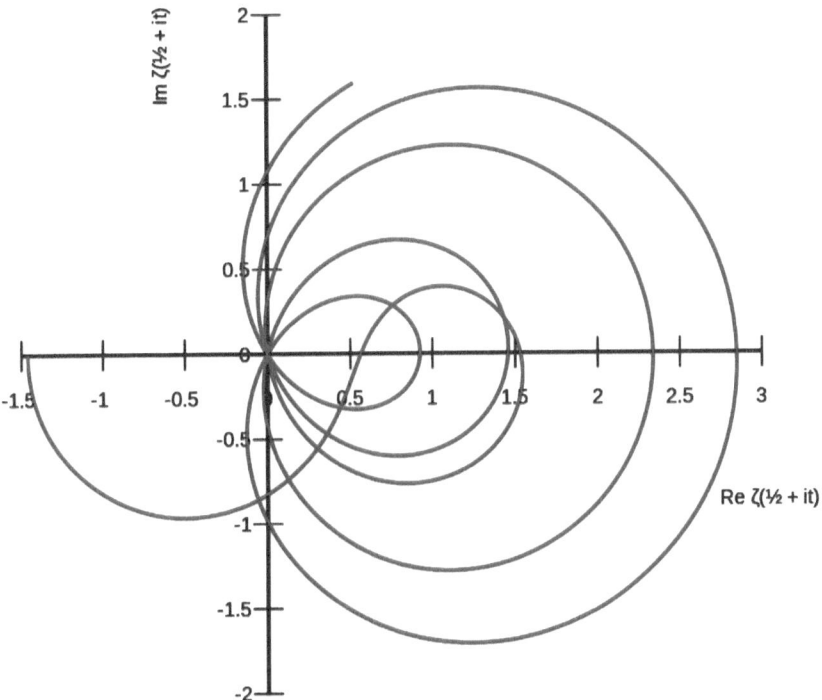

Polar graph of Riemann zeta(½ + it)

The issue of polarity leads us to think of another great theoretical problem for which an experimental correlate is sought: the Riemann zeta function. As we know, basic aspects of this function present an enigmatic similarity with the random matrices of subatomic energy levels and many other collective features of quantum mechanics. Science looks for mathematical structures in physical reality, but here on the contrary we would have a physical structure reflected in a mathematical reality. Great physicists and mathematicians such as Berry or Connes proposed more than ten years ago to confine an electron in two dimensions under electromagnetic fields to "get its confession" in the form of the zeros of the function.

There has been a great deal of discussion about the dynamics capable of recreating the real part of the zeros of the Riemann zeta function. Berry surmises that this dynamics should be irreversible, bounded and unstable, which makes a big difference for the ordinary processes expected from the current view of fundamental fields, but it is closer to quantum thermomechanics, or what is the same, irreversible quantum thermodynamics. Moreover, it seems that what it is at stake here is the most basic arithmetic relationship between addition and multiplication, as opposed to the scope of multiplicative and additive reciprocities in physics.

Most physicists and mathematicians think that there is nothing to scratch about the nature of the imaginary numbers or the complex plane; but as soon as they have to deal with the zeta function, there is hardly anyone who does not begin with an interpretation of the zeros of the function —especially when it comes to its connections with quantum mechanics. So it seems that only when there are no solutions the interpretation matters.

Maybe it would be healthy to purge the calculus in the sense that Mathis asks for and see to what extent it is possible to obtain results in the domain of quantum mechanics without resorting to complex numbers or the stark heuristics methods of renormalization. In fact, in Mathis' version of calculus each point is equivalent to at least one distance, which should give us additional information. If the complex plane allows extensions to any dimension, we should check what is its minimum transposition into real numbers, both for physical and arithmetic problems. After all, Riemann's starting point was the theory of functions, rather than number theory.

Surely if physicists and mathematicians knew the role of the complex plane in their equations they would not be thinking of confining electrons in two dimensions and other equally desperate attempts. The Riemann zeta function is inviting us to inspect the foundations of calculus, the bases of dynamics, and even our models of the point particles and elementary charge.

The zeta function has a pole at unity and a critical line with a value of 1/2 where lie all its known non-trivial zeros. The carriers of "elementary charge", the electron and the proton, both have a spin with a value of 1/2 and the photon that connects them, a spin with a value equal to 1. But why should spin be a statistical feature and not charge? Possibly the interest of the physical analogies for the zeta function would be much greater if the concept of elementary charge were to be dispensed with.

That the imaginary part of the electron wave function is linked to spin and rotation is no mystery. But the imaginary part associated with the quantization of particles of matter or fermions, among which is the electron, has no obvious relation with spin. However, in classical electromagnetic waves we can deduce that the imaginary part of the electrical component is related to the real part of the magnetic component, and vice versa. The scattering amplitudes and their analytical continuation cannot be separated from the spin statistics, and vice versa; and both are associated with timelike and spacelike phenomena respectively. There can be also different analytical continuations with different meanings and geometrical interpretations in the Dirac algebra.

In electrodynamics all the development of the theory goes very explicitly from global to local. Gauss' divergence theorem of integral calculus, which Cauchy used to prove his residue theorem of complex analysis, is the prototype of the cyclic or period integral. Like Gauss law of electrostatics, is originally independent of metric, though this is seldom taken into account. The Aharonov-Bohm integral, a prototype of geometric phase, is very similar in structure to the Gauss integral.

As Evert Jan Post emphasize time and again, the Gauss integral works as a natural counter of net charge, just as the Aharonov-Bohm integral works for quantized units of flux in self-interaction of beams. This clearly speaks in favor of the ensemble interpretation of quantum mechanics, in contrast to the Copenhagen interpretation stating that the wave function corresponds to a single individual system with special or non-classical statistics. The main statistical parameters here would be, in line with Planck's 1912 work in which he introduced the zero point energy, the mutual phase and orientation of the ensemble elements [51]. Of course orientation is a metric-independent property.

Naturally, this integral line of reasoning shows its validity in the quantum Hall effect and its fractional variant, present in two-dimensional electronic systems under special conditions; this would bring us back to the above-mentioned attempts of electron confinement, but from the angle of classical, ordinary statistics. In short, if there is a correlation between this function and atomic energy levels, it should not be attributed to some special property of quantum mechanics but to the large random numbers generated at the microscopic level.

If we cannot understand something in the classical domain, hardly we will see it more clearly under the thick fog of quantum mechanics. There are very significant correlates of Riemann zeta in classical physics without the need to invoke quantum chaos; and in fact well known models such as those of Berry, Keating and Connes are semiclassical, which is a way to stay in between. We can find a purely classical exponent of the zeta function in dynamic billiards, starting with a circular billiard, which can be extended to other shapes like ellipses or the so-called stadium shape, a rectangle capped with two semicircles.

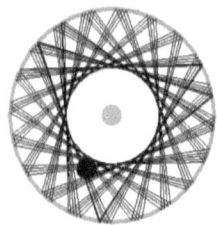
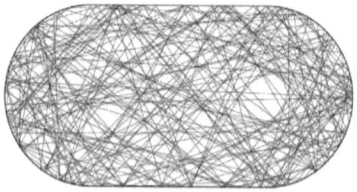

The circular billiard dynamics with a point particle bouncing is integrable and allows to model, for instance, the electromagnetic field in resonators such as microwave and optical cavities. If we open holes along the boundary, so that the ball has a certain probability of escape, we have a dissipation rate and the problem becomes more interesting. Of course, the probability depends on the ratio between the size of the holes and the size of the perimeter.

The connection with the prime numbers occurs through the angles in the trajectories according to a module 2 $\pi/n$. Bunimovich and Detteman showed that for a two hole problem with holes separated by 0°, 60°, 90°, 120° or 180° the probability is uniquely determined by the pole and non-trivial zeros of the Riemann zeta function [52]. The sum over can be determined explicitly and involves terms that can be connected with Fourier harmonic series and hypergeometric series. I do not know if this result may be extended to elliptical boundaries, which are also integrable. But if the boundary of the circle is smoothly deformed to create a stadium-like shape, the system is not integrable anymore and becomes chaotic, requiring an evolutionary operator.

Dynamic billiards have multiple applications in physics and are used as paradigmatic examples of conservative systems and Hamiltonian mechanics; however the "leakage" allowed to the particles is nothing more than a dissipation rate, and in the most explicit manner possible. Therefore, our guess that the zeta should be associated with thermomechanical systems that are irreversible at a fundamental level it is plain to understand —it is Hamiltonian mechanics that begs the exception to begin with. Since it is assumed that fundamental physics is conservative, there is a need for "a little bit open" closed systems, when, according to our point of view, what we always have is open systems that are partially closed. And according to our interpretation, physics has the question upside-down: a system is reversible *because* it is open, and it is irreversible to the extent that becomes closed, differentiated from the fundamental homogeneous medium.

From a different angle, the most apparent aspect of electromagnetism is light and color. Goethe said that color was the frontier phenomenon between shadow and light, in that uncertain region we call penumbra. Of course his was not a physical theory, but a phenomenology, which not only does not detract from it but adds to it. Schrödinger's theory of color space of 1920, based on

[103]

arguments of affine geometry though with a Riemannian metric, is somewhat halfway between perception and quantification and can serve us, with some improvements introduced later, to bring together visions that seem totally unconnected. Mazilu shows that it possible to obtain matrices similar to those arising from field interactions [53].

Needless to say, the objection to Goethe was that his concept of polarity between light and darkness, as opposed to the sound polarity of electric charge, was not justified by anything, but we believe that it is just the opposite. All that exists are density gradients; light and shadow can create them, a charge that is only a + or — sign attached to a point, can not. Colors are within light-and-darkness as light-and-darkness are within space-and-matter.

It is said, for example, that the Riemann zeta function could play the same role for chaotic quantum systems as the harmonic oscillator does for integrable quantum systems. Maybe. But do we know everything about the harmonic oscillator? Far from it, as the Hopf fibration or the monopole remind us. On the other hand, the first appearance of the zeta function in physics was with Planck's formula for blackbody radiation, where it enters the calculation of the average energy of what would later be called "photons".

The physical interpretation of the zeta function always forces us to consider the correspondences between quantum and classical mechanics. Therefore, a problem almost as intriguing as this should be finding a classical counterpart to the spectrum described by Planck's law; however, there seems to be no apparent interest in this. Mazilu, again, reminds us the discovery by Irwin Priest in 1919 of a simple and enigmatic transformation under which Planck's formula yields a Gaussian, normal distribution with an exquisite correspondence over the whole frequency range [54].

In fact, the correspondence between quantum mechanics and classical mechanics is an incomparably more relevant issue on a practical and technological level than that of the zeta function. As for the theory, there is no clarification from quantum mechanics about where the transition zone would be. There are probably good reasons why such a sensitive point receives apparently so little attention. However, it is highly probable that Riemann's zeta function unexpectedly connects different realm of physics, making it a mathematical object of unparalleled depth.

The point is that there is no interpretation for the cubic root of the frequency of Priest's transformation. Referring to the cosmic background radiation, "and perhaps to thermal radiation in general", Mazilu tries his best guess in order to finish with this observation: "Mention should be made that such a result can very well be specific to the way the measurements of the radiation are usually made, i.e. by a bolometer". Leaving this transformation aside, various more or less direct ways of deriving Planck's law from classical assumptions have been proposed, from that suggested by Noskov on the basis of Weber's electrodynamics to that of C.K. Thornhill [55]. Thornhill proposed a kinetic theory of electromagnetic radiation with a gaseous ether composed of an infinite variety of particles, where the frequency of electromagnetic waves is correlated with the energy per unit mass of the particles instead of just the energy —obtaining the Planck distribution in a much simpler way.

The statistical explanation of Plank's law is already known, but Priest's Gaussian transformation demands a physical explanation for classical statistics. Mazilu makes specific mention of the measurement device used, in this case the bolometer, based on an absorptive element —the Riemann zeta function is said to correspond to an absorption spectrum, not an emission spectrum. If today metamaterials are used to "illustrate" space-time variations and mock black holes —where the zeta function is also used to regularize calculations- they could be used with much sounder arguments to study variations of the absorption spectrum to attempt a reconstruction by a sort of reverse engineering. The enigma of Priest's formula could be approached from a theoretical as well as a practical and constructive point of view —although the very explanations for the performance of metamaterials are controversial and would have to be purged of numerous assumptions.

Of course by the time Priest published his article the idea of quantization had already won the day. His work fell into oblivion, and little else is known about the author except for his dedication to colorimetry, in which he established a criterion of reciprocal temperatures for the minimum difference in color perception [56]; of course these are perceptive temperatures, not physical ones; the correspondence between temperatures and colors was never found. If there is an elementary additive and subtractive algebra of colors, there must also be a product algebra, surely related to their perception. This brings to mind the so-called non-spectral *line of purples* between the red and violet ends of the spectrum, the perception of which is limited by the luminosity function. With a proper correspondence this could give as a beautiful analogy that the reader may try to guess.

On the other hand, it should not be forgotten that Planck's constant has nothing to do with the uncertainty in the energy of a photon, though today they are routinely associated [57]. Noskov's longitudinal vibrations in the moving bodies immediately remind us of the *zitterbewegung* or "trembling motion" introduced by Schrödinger to interpret the interference of positive and negative energy in the relativistic Dirac equation. Schrödinger and Dirac conceived this motion as "a circulation of charge" generating the magnetic moment of the electron. The frequency of the rotation of the zbw would be of the order of $10^{21}$ hertz, too high for detection except for the resonances.

David Hestenes has analyzed various aspects of the *zitter* in his view of quantum mechanics as self-interaction. P. Catillon et al. conducted an electron channeling experiment on a crystal in 2008 to confirm de Broglie's internal clock hypothesis. The resonance detected experimentally is very close to the de Broglie frequency, which is half the frequency of the *zitter*; the de Broglie period would be directly associated with mass, as has recently been suggested. There are several models for creating resonances with the electron reproducing the zeta function in cavities and Artin's dynamic billiards, but they are not usually associated with the zitter or the de Broglie internal clock, since this does not fit into the conventional version of quantum mechanics. On the other hand, it would be advisable to consider a totally classical zero-point energy as can be followed from the works of Planck and his continuators in stochastic electrodynamics, though all these models are based *on point particles* [58].

Math, it is said, is the queen of sciences, and arithmetic the queen of mathematics. The fundamental theorem of arithmetic places prime numbers at its center, as the more irreducible aspect of the integer numbers. The main problem with prime numbers is their distribution, and the best approximation to their distribution comes from the Riemann zeta function. This in turn has a critical condition, which is precisely to find out if all the non-trivial zeros of the function lie on the critical line. Time past and competition among mathematicians have turn the task of proving the Riemann hypothesis into the K-1 of mathematics, which would involve a sort of duel between man and infinity.

William Clifford said that geometry was the mother of all sciences, and that one should enter it bending down like children; it seems on the contrary that arithmetic makes us more haughty, because in order to count we do not need to look down. And that, looking down to the most elementary and forgetting about the hypothesis as much as possible, would be the best thing for the understanding of this subject. Naturally, this could also be said of countless questions where the overuse of mathematics creates too rarefied a context, but here at least a basic lack in understanding is admitted.

There seem to be two basic ways of understanding the Riemann zeta function: as a problem posed by infinity or as a problem posed by unity. So far modern science, driven by the history of calculus, has been much more concerned with the first aspect than with the second, even if both are inextricably linked.

It has been said that if a zero is found outside the critical line —if Riemann's hypothesis turns out to be false-, that would create havoc in number theory. But if the first zeros already evaluated by the German mathematician are well calculated, the hypothesis can be practically taken for granted, without the need to calculate more trillions or quadrillions of them. In fact, and in line with what was said in the previous chapter, it seems much more likely to find flaws in the foundations of calculus and its results than to find zeros off the line, and in addition the healthy creative chaos that would produce would surely not be confined to a single branch of math.

Of course, this applies to the evaluation of the zeta function itself. If Mathis' simplified calculus, using a unitary interval criterion, finds divergences even for the values of the elementary logarithmic function, these divergences would have to be far more important in such convoluted calculations like those of this function. And in any case it gives us a different criterion for the evaluation of the function; furthermore, there might be a criterion to settle if certain divergences and error terms cancel out.

The devil's advocates in this case would not have done the most important part of their work yet. On the other hand, fractional derivatives of this function have been calculated allowing us to see where the real and imaginary parts converge; this is of interest both for complex analysis and physics. In fact it is known that in physical models the evolution of the system with respect to the pole and zeros usually depends on the dimension, which in many cases is fractional or fractal, and even multi-fractal for potentials associated with the number themselves.

Arithmetic and counting exist primarily in the time domain, and there are good reasons to think that methods based on finite differences should take a certain kind of preference when dealing with changes in the time domain —since with infinitesimals the act of counting dissolves. The fractional analysis of the function should also be concerned with sequential time. Finally, the relationship between discrete and continuous variables characteristic of quantum mechanics should also be connected with the methods of finite differences.

Quantum physics can be described more intuitively with a combination of geometric algebra and fractional calculus for cases containing intermediate domains. In fact, these intermediate domains can be much more numerous than we think if we take into account both the mixed assignment of variables in orbital dynamics and the different scales at which waves and vortices can occur between the field and the particles in a different perspective like Venis'. The same self-interaction of the *zitterbewegung* calls for a much greater concreteness than hitherto achieved. This movement allows, among other things, a more directly geometric, and even classical, translation of the non-commutative aspects of quantum mechanics which in turn allow for a key natural connection between discrete and continuous variables.

Michel Riguidel makes the zeta function object of an intensive work of interaction in search of a morphogenetic approach. It would be great if the computing power of machines can be used to refine our intuition, interpretation and reflection, rather than the other way around. However, here it is easy to present two major objections. First, the huge plasticity of the function, which although completely differentiable, according to Voronin's theorem of universality contains any amount of information an infinite number of times.

The second objection is that if the function already has an huge plasticity, and on the other hand graphics can only represent partial aspects of the function at any rate, further deformations and transformations, however evocative they may be, still introduce new degrees of arbitrariness. The logarithm can be transformed into a spiral halfway between the line and the circle, and create spiral waves and whatnot, but in the end they are just representations. The interest, at any rate, is in the interaction function-subject-representation —the interaction between mathematical, conceptual and representational tools.

But there is no need for more convoluted concepts. The greatest obstacle to go deeper into this subject, as in so many others, lies in the stark opposition to examining the foundations of calculus, classical and quantum mechanics. The more complex the arguments to prove or disprove the hypothesis, be it true or false, the less importance the result can have for the real world.

It is often said that the meaning of the Riemann hypothesis, and even of all the computed zeros of the function, is that the prime numbers have as random a distribution as possible, which of course leaves wide open how much randomness is possible. We may have no choice but to talk about apparent randomness.

But even so, there we have it: the highest degree of apparent randomness in a simple linear sequence generalizable to any dimension hides an ordered structure of unfathomable richness.

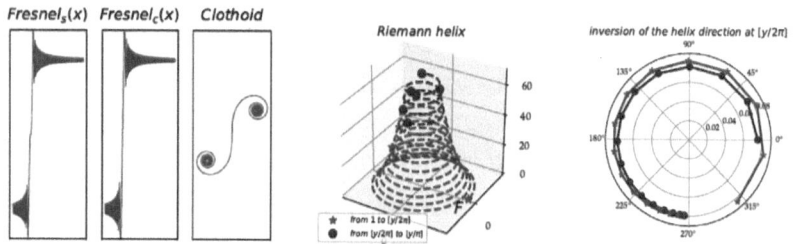

Michel Riguidel: *Morphogenesis of the Zeta Function in the Critical Strip by Computational Approach*

\*

Let us return to the qualitative aspect of polarity and its problematic relationship with the quantitative realm. Not only the relation between the qualitative and the quantitative is problematic, but the qualitative interpretation itself raises a basic question inevitably connected with the quantitative.

For P. A. Venis everything can be explained with yin and yang, seen in terms of expansion and contraction, and of a higher or a lower dimension. Although this interpretation greatly deepens the possibility of connection with physics and mathematics, the version of the yin yang theory he uses is that of the Japanese practical philosopher George Ohsawa. In the Chinese tradition, yin is basically related to contraction and yang to expansion. Venis surmises that the Chinese interpretation may be more metaphysical and Ohsawa's more physical; and latter he thinks that the former

[110]

could be more related to the microcyclical processes of matter and the latter to the mesocyclical processes more typical of our scale of observation, but both views seem to be quite divergent.

Without resolving these very basic differences we cannot expect to soundly connect these categories with quantitative aspects, although one may still speak of contraction and expansion, with or without relation to dimensions. But on the other hand, any reduction of such vast and nuanced categories to mere linear relations with coefficients of separate aspects such as "expansion" or "contraction" runs the risk of becoming a poor simplification dissolving the value of the qualitative in appreciating nuances and degrees.

Venis' interpretation is not superficial at all, and on the contrary it is easy to see that it gives a much deeper dimension, quite literally, to these terms. The extrapolation to aspects such as heat and color may seem to lack the desirable quantitative and theoretical justification, but in any case they are logical and consistent with his general vision and are wide open to delve into the subject. However, the radical disagreement on the most basic qualifications is already a challenge for interpretation.

It should be said right from the start that the Chinese version cannot be reduced to the understanding of yin and yang as contraction and expansion, nor to any pair of conceptual opposites to the exclusion of all the others. Contraction and expansion are only one of the many possible pairs, and even if they are often used, as with any other pair they depend entirely on the context. Perhaps the most common use is that of "full" and "empty", which on the other hand is intimately linked to contraction and expansion, although they are far from identical. Or also, depending on the context, the *tendency* towards fullness or void; it is not for nothing the common distinction between young and old yang, or young and old yin. Reversal is the way of Tao, so it is only natural that these points of potential, spontaneous reversal are also notorious in the *Taijitu*.

On the other hand, qualities such as full and empty not only have a clear translation in differential terms for field theories, hydrodynamics or even thermodynamics, but also have an immediate meaning, although much more diffuse, for our inner sense, which is precisely the common sense or sensorium as a whole, our undifferentiated sensation prior to the imprecise "sensory cut" which seems to generate the field of our five senses. This common sensorium also includes kinesthesia, our immediate perception of movement and our self-perception, which can be both of the body and of consciousness itself.

This inner sense or common sensorium is just another expression for the homogeneous, undivided medium that we already are —the background and tacit reference for feeling, perception and thought. And any kind of intuitive or qualitative knowledge takes that as the reference, which obviously goes beyond any rational or sensory criteria of discernment. Conversely, we could say that this background is obviated in formal thought but is assumed in intuitive knowledge. Physicists often speak of a result being "counter-intuitive" only in the sense that it goes against the expected or acquired knowledge, not against intuition, which it would be vain to define.

However it would be absurd to say that the qualitative and the quantitative are completely separate spheres. Mathematics is both qualitative and quantitative. We use to hear that there are more qualitative branches, like topology, and more quantitative branches like arithmetic or calculus, but on closer inspection this hardly makes any sense. Venis' morphology is totally based on the idea of flow and on such elementary notions as points of equilibrium and points of inversion. Newton himself called his differential calculus "method of fluxions", the analysis of fluent quantities, and the methods for evaluating curves are based on the identification of turning points. So there is a compatibility that not only is not forced but it is natural; if modern science has advanced in the opposite direction towards increasing abstraction, which in turn is the just counterbalance to its utilitarianism, is another story.

Polarity and duality are quite different things but it is useful to perceive their relation before the convention of electric charge was introduced. The reference here cannot fail to be the electromagnetic theory, which is the basic theory of light and matter, and to a large extent also of space and matter.

Obviously, it would be absurd to say that a positive charge is yang and a negative charge is yin, since between both there is only an arbitrary change of sign. In the case of an electron or a proton other factors come in play, such as the fact that one is much less massive than the other, or that one is peripheral and the other is at the core of the atom. Let us take another example. At the biological and psychological level, we live between stress and pressure, which frame our way of perceiving things. But it would also be absurd to say that one or the other is yin or yang insofar as we understand tension only as a negative pressure, or viceversa. In other words, mere changes of sign seems to us trivial; but they become more interesting qualitatively and quantitatively when they involve other transformations.

Whether all or nothing is trivial depends only on our knowledge and attention; a superficial knowledge may judge as trivial things that in fact are are full of content. The polarity of charge may seem trivial, as may seem the duality of electricity and magnetism, or the relationship between kinetic and potential energy. Actually none of them is trivial at all, but when we try to see everything together we already have a space-time algebra with a huge range of variants.

In the case of pressure and stress or tension, the more apparent transformation is the deformation of a material. Strain-stress-pressure relations define, for instance, the properties of the pulse, whether in the pulsology of traditional Chinese or Indian medicine or in modern quantitative pulse analysis; but that also leads us to the stress-strain relations that define the constitutive law in materials science. Constitutive relations, on the other hand, are the complementary aspect of Maxwell's electromagnetic field equations that tell us how the field interacts with matter.

It is usually said that electricity and magnetism, which are measured with dimensionally different units, are the dual expression of the same force. As we have already pointed out, this duality implies the space-matter relationship, both for the waves and for what is supposed to be the material support of the electric and magnetic polarity; in fact, and without going into further detail, this seems to be the key distinction.

All gauge field theories can be expressed by forces and potentials but also by non-trivial pressure-strain-stress variations that involve feedback, and there is feedback at any rate because first of all there is a global balance, and only then a local one. These relations are already present in Weber force law, only in this one what is "deformed" is force, instead of matter. The great virtue of Maxwell's theory is to make explicit the duality between electricity and magnetism, hidden in Weber's law. But we must insist, with Nicolae Mazilu, that we can find the essence of the gauge theory already in Kepler's problem.

Constitutive relations with definite values such as permittivity and permeability cannot occur in empty space, so they can only be a statistical average of what occurs in between matter and space. Matter can sustain stress without exhibiting strain or deformation, and space can deform without stress or tension —this runs in parallel with the basic signatures of electricity and magnetism, which are tension and deformation. Strain and tension are not yin or yang, but to yield easily to deformation is yin, and to withstand tension without deformation is yang —at least as far as the material aspect is concerned. Of course between both there must be a whole continuous spectrum, often affected by many other considerations.

However, from the point of view of space, to which we do not have direct access but through the mediation of light, we could see the opposite: the expansion without coercion would be pure yang, while the contraction may be seen as a reaction of matter to the expansion of space, or the radiations that fill it. The waves of radiation themselves are an intermediate and alternate process between contraction and expansion, between matter and space, which cannot exist separately. However, a deformation is a purely geometrical concept, while a tension or a force is not, being here where the proper domain of physics begins.

[114]

Perhaps in this way a criterion for reconciling the two interpretations can be discerned, not without careful attention to the overall picture of which they are part; each may have its range of application, but they cannot be totally separate.

It is a law of thought that concepts appear as pairs of opposites, there being an infinity of them; finding their relevance in nature is something else, and the problem becomes nearly unsolvable when quantitative sciences introduce their own concepts that are also subject to antinomies but of a very different order and certainly much more specialized. However, the simultaneous attention to the whole and to the details makes this a task far from impossible.

Much have been said about holism and reductionism in sciences but it must be remembered that physics to start with, never has been described in rigorously mechanical terms. Physicists hold onto the local application of gauge fields, only because that is what give them predictions, but the very concept of Lagrangian that makes all that possible is integral or global, not local. What is surprising is that this global character has not a proper use in fields such as medicine or biophysics.

Starting from these global aspects of physics, a genuine and meaningful connection between the qualitative and the quantitative is much more feasible. The conception of yin and yang is only one of many qualitative readings man has made of nature, but even taking into account the extremely fluid character of these distinctions it is not difficult to establish the correspondences. For example, with the three gunas of Samkya or the four elements and four humors of the Western tradition, in which fire and water are the extreme elements and air and earth are the intermediate ones; these also can be seen in terms of contraction and expansion, of pressure, tension and deformation.

Needless to say, the idea of balance is not exclusive of the Chinese conception either, since the same cross and the quaternary have always had a connotation of equilibrium that is totally elemental and of universal character. It is rather in modern physics that equilibrium ceases to have a central place due to inertia, although it cannot cease to be omnipresent and essential for the use of reason itself, as it is for logic and algebra. The possibility of contact between quantitative and qualitative knowledge depends both on the precise location we give to the concept of equilibrium and the correct appreciation of the context and global features of the so-called mechanics.

Unlike the usual scientific concepts, which inevitably tend to become more detailed and specialized, notions such as yin and yang are ideas of utmost generality, indexes to be identified in the most specific contexts; if we pretend to define them too much they also lose the generality that gives them their value as an intuitive aid. But also the most general ideas of physics have been subject to constant evolution and modification depending on the context, and we only have to look at the continuous transformations of quantitative concepts such as force, energy or entropy, not to mention issues such as the criterion and range of application of the three principles of classical mechanics.

Vortices can be expressed in the elegant language of the continuum, of compact exterior differential forms or geometric algebra; but vortices speak above all with a language very similar to that of our own imagination and the plastic imagination of nature. Therefore, when we observe the Venis sequence and its wide range of variations, we know that we have find an intermediate, but genuine, ground between mathematical physics and biology. In both, form follows function, but in the reverse engineering of nature that human science is, function should follow form to the very end.

In Venis account there is a dynamic equilibrium between the dimensions in which the vortex evolves. This widens the scope of the equilibrium concept but makes it more problematic to assess. Fractional calculus would have to be key to follow this evolution through the intermediate domains between dimensions, but this also rise interesting points for experimental measurements.

How dimensions higher than three can be interpreted is always an open question. If instead of thinking of matter as moving in a passive space, we think of matter as those portions to which space has no access, the same matter would start from the point or zero dimension. Then the six dimensions of the evolution of vortices would form a cycle from the emission of light by matter to the retraction of space and light into matter again —and the three additional dimensions would only be the process in the opposite direction, and from an inverse optic, which circumvent repetition.

This is just one way of looking at some aspects of the sequence among many possible ways, and the subject deserves a much more detailed study than we can devote to it here. One think is to look for some sort of symmetry, but there must be many more types of vortices than we know now, not to speak of the different scales of occurrence, and the multiple metamorphoses. Only in Venis' work one can find the due introduction to these questions. Venis assumes the number of dimensions to be infinite, so we could not find and count them all. An indication of this would be the minimum number of meridians necessary to create a vortex, which increases exponentially with the number of dimensions and which the author associates with the Fibonacci series.

*We can speak of polarity as long as we can appreciate a capacity for self-regulation.* That is to say, not when we just count on apparently antagonistic forces, but when we can not help notice a principle above them —or an underlying unity, if preferred. This capacity was always present since the very Kepler's problem, and it is only telling that science has failed to recognize it. Weber's force and potential are explicitly polar, Newton's force is not, but the two-body problem exhibit a polar dynamics in any case. To call the evolution of celestial bodies "mechanics" is just a rationalization, and in fact we do not have a mechanical explanation of anything when we speak of fundamental forces, and probably we cannot have one. Only when we notice a self-regulating principle could we use the term *dynamics* honoring the original intention still present in that name.

## 11. Time and the Clock

Pendulums, circles, waves, vortices, electrons, ellipses, spirals... What kind of "clock" could be built with all this? Or better yet, what kind of "time" would it be measuring, counting, calculating? Furthermore, would it be calculating it, or rather indicating it?

To speak of the third principle of dynamics is to speak of the very concept of reciprocity —within closed systems. The reciprocity of relativistic reference frames is purely mathematical, since it is not bound to centers of mass —but the same applies to forces in ellipses according to Newton.

Relativistic simultaneity, even the local determinism of quantum mechanics, is embedded in the basic Newtonian assumption of the global synchronizer, the absolute time in which the third principle does not take place sequentially but simultaneously. But the same vortex of Newton's bucket experiment, as Pinheiro notes, denies this absolute time in the most convincing and categorical way —and not only for instinct, which is always stronger than metaphysics. The thermomechanical reading allows us to obtain more information, as well as another type of indication. And that same vortex is already a completely different model of clock, that we should learn to contemplate —if to contemplate something, in this our artificial world, we have to start by being able to somehow reproduce it.

We agree here with Pinheiro and other authors [59]. Various arguments and numerical calculations indicate that the third principle is not applicable to systems out of equilibrium; and if it is applicable, it can only be so in the sense of the retarded potentials. That is, we assume that even celestial orbits are always out of equilibrium and only reach it instantly thanks to an additional term representing the action of the medium on matter; this term can be entropic, or respond to a longitudinal vibration, or a constant bombardment of external particles.

In quantum mechanics, which is supposed to be the fundamental theory, forces become secondary to the potential; however, we continue to understand everything according to the logic of the three principles, even after the relativistic generalization of the principle of equivalence.

But the decisive turn had already taken place when the potential went from representing a mere position to having an irreducible temporal component. For Weber's relational dynamics that emerged halfway between Newtonian and 20th century physics, it does not make sense to separate the dynamic component of the state of the system, nor the forces of the environment in which they occur. The rigorous application of the principle of dynamic equilibrium also made this distinction irrelevant, as it made irrelevant the equivalence principle, since inertia itself is dispensable.

If, by virtue of the principle of dynamic equilibrium, we dispense with the concept of inertia, we also dispense with the intentionality —the *dispositio* underlying mechanics. The Global Synchronizer, so perfectly intangible, is the supreme symbol of power in the present civilizing cycle. It is intangible as it is metaphysical, but it suppresses or in any case makes unrecognizable the local exchange of information of the open systems and the interaction with the environment.

It has been said that after Newton the universe went from breathing like a mother to ticking like a clock; there is still time to bring it back to breath again.

## 12. Tao of Technoscience

The paths in the science-technology continuum may be innumerable but they all presuppose a potential reciprocity between knowledge and application —thus between knowledge and power. And yet we still have no idea of what kind of circle knowledge and power draws on us.

Newton's celestial mechanics seemed initially far removed from worldly affairs but the unwarranted generalization of his principles to things far removed from human artifacts had the effect of turning the world into a wheelless rolling machine.

Society has taken shape as it becomes isolated from Nature but cannot subsist without a permanent commerce with her which in turn depends more and more on our knowledge of it. Any dominance relationship over Nature is reproduced within society, between some parts that exercise control and others parts subject to this control.

The solar system bound by gravity, or the function of the heart in our blood circulation, have been seen as simply governed by the concept of force in our present world view. Since the middle of the 20th century, stability theory and cybernetics developed a theory of control over these so-called "blind forces", generalizing a version of entropy that was already far removed from the original thermodynamical context. Now it remains to be seen what twist would result for control theory assuming spontaneous regulation in action principles, in the Second Law and in the collective resonance between elements; as well as in the relationship between these three aspects.

Overcoming reductionism cannot in any way involve last-minute corrections that seek to compensate increasing degrees of abstraction with also increasing degrees of subjectivity for the sake of the inclusion of the observer, be it in statistical mechanics, quantum mechanics or relativity; it involves in any case correcting the gaps in the foundational position, which so far remains unaffected.

But the reciprocity between man and nature goes far beyond anything we suspect, and cannot be encompassed by a mere theoretical turn, however wide or deep it may seem. It is not a matter of looking for an idealized external nature either, since all that is trapped in the human being is also nature.

We do not know and maybe we do not want to live without machines. Can we radically change our relationship with them? Machines, too, are trapped, molded and compressed nature; and while we are forced to depend on them we are routinely trained for obedience. I will now pick up a few paragraphs from a topic I discussed at greater length in *Techno-Science and the Laboratory of Self*:

"Vico's principle, which states that knowing is making, is more general than Descartes'. But surely one can also doubt Vico's principle. I can move my hand, but do I know how I move my hand? Second hand, so to speak, not first hand. Of course making is not doing, except in thinking, and we make machines not to do things directly, and not to directly think. We can then try to introduce into the realm of knowledge-power the duly reformed Vico principle: I only know that in what I take part, and to the extent that I take part.

It is not by calculus, but by the practical arts, that we know the world best. The same concept of efficiency, as economy of effort or elegance, was a natural notion in the art of all cultures before techniques were invaded by stacked layers of scientific mediations;

now it would have to be taken out of the bottom of the stack. There is a natural sense of efficiency in any physical activity, in the right intonation, in any gesture or brushstroke.

To move from one area to another, from the functional domain governed by calculus to the intuitive functioning, we can take as example the biological feedback and biofeedback. A signal that corresponds to a vital function can be used to vary it at will, within certain limits of course. However, and this is the important thing, here any notion of manipulation is out of place, as in this context loses all meaning. Even control, with all its vast current theory, is subsumed in the idea of self-control, which far from being a particular case, seems to be the most indefinite and general.

In our ordinary physical control of external objects, the relationship between action and calculus is also reversed. Think of the complicated balance involved riding a bicycle; dynamics can hardly solve the problem by means of centrifugal forces in the case of slow motion, but it gets out of hands in cases with higher speeds. And yet for the cyclist it is just the opposite: speed is the solution, and excessive slowness the problem. Motion is shown by pedalling.

However, within the category of self-control there is more than just cycles of perception and action; there is also self-observation. In the case of biofeedback there are two basic cases, direct monitoring of a function, such as when we observe our breathing and modify it without even intending to do so, and indirect monitoring, either by means of a mirror that returns our image to try to move involuntary muscles or by a device with sensors that translates signals generated by ourselves without being aware of it.

The biofeedback motif may seem very limited as it has hardly transcended the level of a curiosity since its appearance and diffusion fifty years ago. However, it marks a turning point in the relationship between man and the machine. If the most helpful idea in explaining the emergence of tools is that they are extensions or prostheses that project our capacity as organisms, and if we have later recognized that from certain point onwards all harmonious relations between the tool and the organ are lost, here for the first time we use the machine to help us regain consciousness of organic functions that have already sunk below the threshold of attention.

So, if technology came out of the biology of the conscious organism, it is precisely here that it returns to it in the most mediated way possible, and with a somewhat undecided intention. Clearly, the entire cybernetic theory of control would have to return to self-control as its archetype, since this one already incorporates the cycles of perception and action, allowing the right space for the self-consciousness around which they revolve; the automatic is subsumed in self-control, self-control in self-interaction and this in spontaneity.

In spite that gauge fields contain a basic feedback mechanism, the use of the Lagrangian in control theory seems totally secondary, which is a curious situation. Things could change working with Weber-type forces, and also with dissipative, thermomechanical forces, in the sense proposed by Pinheiro. The measure theory would also have to be adjusted to the different requirements of this type of systems.

Today it is said that the double access to perception and action defines an artificial intelligence problem. But a greater field in self-perception and natural intelligence awaits us.

## 13. Individual evolution of an entity

The so-called "new synthesis" of the theory of evolution has never had the least predictive power, but neither has any descriptive power, and like cosmology, it has been used mostly as a narrative complement for normative physical laws unable to connect with the real world and the real time in the most decisive sense.

To alleviate the evident limitations of a theory that only through phantasy can have some contact with natural forms —and let us not underestimate the power of fantasy in this instance-, evolution has been combined with the perspective of biological development (evo-devo), and even with ecology (eco-evo-devo), but even then it has not been possible to create a moderately unitary framework to describe problems such as the emergence, maturation, aging and death of organized beings.

There is no need to call all this by any other name than evolution without further ado, since the present theory has only taken over the name to deal with speculative and remote issues, rather than with close and fully approachable problems.

What determines the aging and death of individual organisms and societies? This is a real evolution problem that takes us closer to the real time.

Astrophysicist Eric Chaisson has observed that the energy rate density is a much more decisive and unequivocal measure for complexity metrics and evolution than the various uses of the concept of entropy, and his arguments are fairly straightforward and convincing [60]. In any case it would be desirable to include this quantity in a context better articulated with other physical principles.

Georgiev et al. have attempted to do so by establishing a feedback loop between this energy flow, the physical principle of minimum action *understood as efficiency or as movement along paths with less curvature or restriction*, and a quantitative principle of maximum action; that is, with the "metabolic" energy flow per mass mediating between efficiency and size, applying it experimentally to CPUs as organized flow systems [61]. When these flow-efficiency-size vertices are connected in a positive feedback loop, an exponential growth of all three is produced and power-law relations between them arise. Although this model admits many improvements and can be illustrated in very different ways, it seems on the right track.

Since the basic problem of aging is the increasing restriction and the inability to overcome it, and any theory not addressing this fundamental issue cannot make a dent in the subject. Another way of saying the same is that organic aging is the increasing inability to eliminate. Aging is irreversibility, and irreversibility is the increasing incapacity to be an open system.

Let us think a bit about this. Something as basic and elementary in physics as the principle of least action is capable of telling us something absolutely essential about aging: to understand the real value of this we only have to know how to apply global measurements in the context of open systems with a variable use of the available free energy.

The theoretical advance in this field is infinitely more feasible than in the modern synthesis and infinitely more relevant, since here it is no longer a question of species, but of the destiny and individual evolution of any spontaneous organization, be it a vortex or a soliton, a human being or a civilization; to a large extent it affects even the Lamarckian evolution of machines and computers with a definite design and purpose.

The energy rate density is a measurable, unambiguous quantity, profoundly significant from a cosmic perspective, and can also have a bearing on bringing down to earth unmanageable entropy criteria. Naturally, the flow-curvature-size criterion can be contrasted with the flow-curvature-entropy criterion, whether the latter is maximum or not.

This threefold flow density-curvature-size criterion can be applied fruitfully to contexts where flow is the decisive factor, be it in strongly quantitative models like the monetary flows, or in purely qualitative models of vortices evolving between expansion and contraction such as the one proposed by Venis. It can be applied even to the circle of individual destiny, suggesting a clock of its evolution, which in today's terminology many would call an aging clock.

## 14. From the Book of Changes to the Algebra of Conscience, through technical analysis

In his splendid book on Harmony Mathematics, Alexey Stakhov devotes a section to Vladimir Lefebvre's mathematical theory of decision and strategic interaction, which, unlike other developments in this area such as the much-hyped game theory, includes an intrinsic behavioral and moral component [62].

Everything starts from the observation that if someone is asked to divide a pile of string beans into two piles of bad and good ones, oddly enough the average result is not 50%-50% as expected, but 62%-38% in favor of those considered good.

According to Lefebvre, what lies at the bottom of this asymmetric perception is a triple gradation of definable logical implications within a binary logic or Boolean algebra:

$$A [a_0 consciousness \rightarrow a_1 reflection \rightarrow a_2 intention]$$

consciousness being the primary field, reflection the secondary level and intention a reflection of second order.

Within this structure of human reflexion, the personal evaluation of behavior implies, by logical implication, a self-image:

$$A = a_0^{a_1^{a_2}} \quad ; \quad A = (a_2 \rightarrow a_1) \rightarrow a_0.$$

The subsequent truth table gives us an asymmetric sequence of 8 possible values, 5 being a positive estimate and 3 a negative one. Naturally, 8, 5 and 3 are adjacent numbers in the Fibonacci series.

The extraordinary thing about the logical framework of ethical cognition created by Lefebvre is that, inadvertently, it allows a completely modern interpretation of the *Book of Changes* without distorting its sapiential and moral nature. It is not something that we are going to prove here, but given the deep influence of *Yi Jing* in the history of Chinese culture, it would be of utmost interest to look for a rigorous correspondence between both views.

Undoubtedly, Lefebvre's framework is more formalized, and more focused on the image that the individual agent has of himself, than on the situation or critical juncture, as is the case with the *Yi Jing*; in fact, we could say that Lefebvre's formalization is a Boolean and measurable limit for interactions with a low number of individual agents, but even then the correspondence remains intact and can lead to much broader and impersonal degrees of involvement, and therefore, of understanding.

Stakhov devotes another section to Elliott's technical wave analysis, a tool also based on the Fibonacci series. The theoretical scope of Elliott waves is limited at best, even if its fractal-type analysis can be applied discretionally. But here we want to point out a much deeper methodological issue.

Lefebvre's model shows us that these series have, or at least admit, a reflexive component —in fact his work has been qualified as a reflexive theory of social psychology. And that reflective component is precisely what is most missing in Elliott's model. There is also a reflexive theory of the economy and the market, with positive and negative feedback cycles; this theory is not without merit, if we compare it, for example, with the hypothesis of the efficient market, which, even if it were not false, would always be incapable of telling us anything. Moreover, the reflexive theory does not only affect prices, but also the fundamentals of the economy.

Economic reflexivity is a model of self-interaction, as it is also of self-image, and this decisive element in human affairs can never be dispensed with, even if it is impossible to quantify it. But what happens when self-interaction runs through the whole behavior of natural systems? Let us think, for example, of the case of the pulse analysis we have mentioned, a transparent model of self-interaction, which, as a bonus, tends towards golden ratios in rhythm and pressure. But it is clear that in social systems reflexivity passes through the knowledge of an external situation, while in an organic system reflexivity is a pure question of action.

How much room does one type of reflexivity leave for the other? How far can the perception of the markets be manipulated? Today we also see that central banks have an undisguised and unabated intervention in prices, injecting money discretionally and becoming guardians of the financial bubbles. Well, the analysis of the pulse, the transformation of its modalities, and the study of the limits of conscious manipulation through biofeedback are giving us a good match or at least some sort of reference for this elusive problem. And if we want more intensive methods, besides the quantitative implications that all this already have, we can also apply to monetary flows CPUs models like that of the triangular flow-efficiency-size feedback that we commented on earlier.

*

In *The Algebra of Conscience*, Lefebvre made a comparative analysis of the Western and Soviet ethical context within the framework of the Cold War confrontation [63]. The W-system of the United States believed that the compromise between good and evil was bad, and that the confrontation between good and evil was good. The S-system of the Soviet world, on the contrary, believed that the compromise between good and evil was good and the confrontation between good and evil was bad —what Lefebvre does not say, as it goes beyond his formal analysis, is that the desire for confrontation of the W system was not based on any moral idiosyncrasy but on the essential expectation of further expansion; just as the desire of the S system to avoid confrontation was based on the fear of disappearing.

At any rate, from the choices and implications between commitment and confrontation, four basic attitudes emerge, that of the saint, the hero, the philistine and the dissembler: the saint embraces suffering and guilt to the maximum; the philistine wants to diminish suffering but may feel acute guilt; the hero minimizes his guilt but not his suffering; and the dissembler minimizes suffering and guilt.

Maybe we could have created 8 types instead of 4, in consonance or dissonance with 8 other scenarios or natural tropisms, with two more third grade binary implications or a single sixth grade one. Logic and calculus serve to ascend this ladder and understanding begins where calculus ends. Thus, the continuous proportion would show us in a very direct way its role of mediator between the discrete and the continuous, the digital and the analogical.

In reality we have here an unsuspected Centaur, namely, a halfway point between binary formal logic and the dialectical logic as in Hegel. But the proof that this is something genuine, and not a mere theoretical construct, is that it has a built-in asymmetry that is a distinctive feature of Nature. We could call it an asymmetrical logic of implication.

It has been said that Lefebvre's theory was used at the highest levels of negotiation during the collapse of the Soviet Union; and here we will not go into judging what positive or negative role it might have had, and for whom. In any case, its framework only covers two agents, and not the scenario or juncture that may dispose of them. *Yi Jing's* framework can serve not only to see how I perceive myself in the world, but also to evaluate how the world perceives me; in the end, both are parallel illusions. One could say that asymmetric perception is part of the same reality; a double asymmetry better embraces the implications and the unnoticed axis of the dynamics of a situation.

Surely one could make a great parlor game out of these arguments. More interesting, however, is the study of moral or social conscience through the analysis of its implications. And even more interesting is the awareness of how the external situation, for which we are always looking for an image, interpenetrates with the changing image we make of ourselves, forming both a momentary map of our situation and conscience.

One can only guess that Lefebvre knows the text of the *Yi Jing*; however it is evident that the line of his reasoning is completely independent from the Chinese classic and he has arrived at his "positional system" of moral consciousness starting from the Western logical and mathematical tradition. If he had had just a glimpse of the *Taijitu* with the golden section embedded, the immediate association would have shine in his mind. In the coming years we may witness many new associations of this kind induced by the evolving environment.

## 15. Science and conscience

We have tried to approach a single subject from the most diverse angles, so that everyone has some possibility of connecting it in the most direct way with his or her own interests. This is a minimal introduction relying on a bibliography that is not at all exhaustive, but necessary to delve into any of the issues presented.

Of course our subject proper was not the continuous proportion, but reciprocity and self-organization in Nature and beyond Nature. The constant $\varphi$ continues to play a marginal, totally episodic role within modern science, which is mainly focused on calculus.

Thus, our preferential point of view is not so much that of science and objectivity, as that of reciprocity itself, to which I attach more importance; as an attempt to give more importance to the awareness of reality than to science.

And reality itself resembles a parable. We can say that next to unity, sometimes also symbolized by the point, the circle and the constant $\pi$, two other great mathematical constants exist from eternity, $e$ and $\varphi$. The constant $e$, the basis of the natural logarithms and the exponential function, perfectly embodies the analytical exhaustiveness of calculus, as if looking towards the plurality of the world. The constant $\varphi$, the natural algorithm of a Nature that completely ignores calculus, looks at unity without knowing it. And unity itself, if it is really unity, cannot show a preference for either.

Now, it is clear that in modern science the weight of $e$ is infinitely greater than that of $\varphi$, to the point that we could perfectly dispense with $\varphi$ without even noticing it. The number $e$ refers us to continuity and infinite divisibility; $\varphi$ refers us to discrete operations without intended purpose —and it should be understood that infinite divisibility is already a human purpose that can always exceed its operational limits. Hence the need to look back to the contradictory basis of infinitesimal calculus.

As it is known, the number $e$ was first identified by Jacob Bernoulli in 1683 in a problem of compound interest, and is absolutely consubstantial with the modern spirit of calculus with all that it entails. The continuous proportion has undoubtedly a much older past but, in the eyes of the moderns, it is hard to see how it could have played a relevant role in the knowledge of antiquity. Euler's number is at the base of so-called advanced or superior mathematics, while the number closest to the muses cannot ever deny the elemental character of its origin —which is also its greatest charm. Surely this is the reason for its permanent popularity among amateur mathematicians.

But this undeniable circumstance hides another naivety on the part of the spirit of calculus that we should learn to appreciate. It is well known that people like Stevin or Newton still believed that ancient cultures could have had broader knowledge than their contemporaries; if mathematicians like them still dared to think that, it must surely be attributed to the incomparable impact that the legacy of Apollonius and Archimedes —the most advanced mathematicians of antiquity - had at that time.

But this already presupposed a totally biased idea about what was advanced in knowledge, which has been perpetuated until today.

It is said that the amount of knowledge regularly doubles every 15 years since the scientific revolution, which implies that today our knowledge of physics and mathematics is four million times greater than in 1687, when the *Principia* were published. Yet this magnum opus is already an obscure, hard to read treatise that has been and continues to be routinely misinterpreted even by the best experts.

Let us try to understand four million *Principia*. We do not walk on the shoulders of giants, the giants advance on our shoulders, although less and less, as it is easy to understand. And the logic of accumulated capital is that accumulated capital does not take risks. The theory of relativity and other "revolutions" were adopted following the principle of minimum elimination and maximum conservation of capital. The whole exacerbated search for novelty in theoretical physics is nothing but the enforced headlong rush because it is not allowed to really examine the foundations. But the less one eliminates, and the less one renews the foundation, the more inexorably one ages.

Obviously the stratification of knowledge and the branching of specialities follows the logic of continuous interest and accumulated capital and debt.

It is really curious that two constants as ubiquitous as $e$ and $\varphi$, the "two natural fractals", cross so little of each other's paths in a field as unlimited but redundant as mathematics. So curious, that the study of the points of contact and divergence of both constants should be an area of mathematical research in its own right, full of interest for both pure and applied mathematics. If this has not happened yet, is because of the one-sided development of all sciences and mathematics on the side of calculus and prediction, putting the rest of the resources at their service.

It is understandable that there is, among many mathematicians, a typical allergic reaction to the questions raised by the continuous proportion and the mathematics of harmony. They are seen more as a hindrance than as a guide, since they might imply some sort of discrete, constructive limits to an analysis for which no restriction is wanted. Nothing should measure the one who measures.

It seems that at the time of the emergence of writing and the first great cities, the priests kept the standards of measurement in the temples. But Nature is the perfect temple that keeps everything without hiding anything.

We have said that φ seems to be "the natural algorithm" of a Nature that does not care in the least about calculus. But not having any system of calculus and measurement is, in a sense, equivalent to having all of them and surpassing them all; in the same way that nature's lack of intention infinitely surpasses human purposes. So, the value of this branch of mathematics for the theory of measurement and computation should be out of the question. It is a matter of identifying relevant problems in the domain of interdependence.

In just over three centuries we have had at least five great cuts with the constructive and proportional conception of measurement: the infinitesimal calculus and Newton's classical mechanics, Maxwell's theory, set theory, relativity and quantum mechanics.

With each of these successive steps, the problem of measurement has become more and more critical and controversial. But it is absolutely superficial to think that only the latest developments count and that measure theory itself is just an aid to our predictions. This is exactly how this tower of Babel was built.

*

The interest of the continuous proportion would go beyond our modern involvement in complexity and calculus. In this area, new discoveries are always possible even at the elementary level that are so unexpected that we do not even know how to value them. Moreover, its persistent appearance in music —that unconscious arithmetic, as Leibniz said-, in physiological rhythms and anatomical sequences indicates not only an intuitive, but, ultimately, pre-numerical character.

There is here a great open track not only for the archaeology of knowledge, but for the very activation of an ancient knowledge that seems inconceivable to us today. In the *Book of Changes* we may see an asymmetric implication algebra. The six lines of a hexagram correspond to the six directions of space at the ends of its three axes, which oppose an I and its circumstance. But here the symmetry of the coordinates is only the external and passive frame, it is the asymmetry what constitutes the internal dynamic spring in which agent and situation interpenetrate.

The *Book of Changes* is the best possible example of an analogical knowledge that does not depend on calculus, and towards which a certain natural logic of implication converges. What is important is not the 64 cases or the 384 lines, but the plane of synthesis towards which they point.

If mathematical analysis has its so-called complex plane to operate with any number of dimensions or variables, one can equally conceive an implication logic that is capable of reducing any number of variables to an equally intangible plane of synthesis, which we will call the plane of universal synthesis or universal inclusion.

Of course, in this day and age anyone who hears of "universal inclusion" can only think of universal confusion. That is why we have developed the analysis, because we do not believe in knowledge that is not properly formalized. However, the development of formalization has not contributed to greater intelligibility, but rather the opposite. For us, formal knowledge increasingly projects more shadows than light; shadows we know too well are cast by ourselves. Shadows of power over natural or social processes.

The whole philosophy of the West since Descartes is based on the idea of a separate intelligence. That is why the successive cuts that come one after another since Newton do not seem a loss to us, since in this line of logic, more separation from Nature amounts to more self-affirmation. But any process has its limit, and when there is nothing left to assert oneself about —because Nature vanished- everything stops making sense, except for the mere exercise of power, which lacks intrinsic stimulus for knowledge.

Besides, the golden age when the field of fundamental predictions was growing is now far behind us. It will not return, for the simple reason that everything was optimized for predictions and all the low hanging fruit is already picked. One can only scrape out diminishing returns in the ugly struggle with complexity. Yet modern science finds it almost impossible to examine its foundations, which is the only thing where there could be real novelty; we have the great advantage of our historical perspective, but the very existence of the specialities depends on not reflecting on foundations.

Modern science is not capable of dealing with either transcendence or immanence; for as subjects we cannot separate ourselves anymore, nor does we know how to see things again from within Nature.

Now, if we look back as we have done, only reordering our perception of the present theories, what does it mean to dispense with the principle of inertia? What is the point of saying that everything is based on self-interaction? What is the point of saying that the only thing we perceive is the Ether?

These are transcendental statements, in the sense that the father of phenomenology, Edmund Husserl might have given them, had he dealt with physics. However, to suspend the principle of inertia ends with the false idealism of physics, such a fecund contradiction that isolates a ball that rolls from the rest of the world except from ourselves. To realize that we only perceive the Ether, because we only perceive in the mode of light, is to realize that we are always right in the middle and that matter and space themselves are transcendental limits.

Finally, to say that the planets or the electrons orbit their centers by self-interaction can only be understood in the sense that the relationship between matter and the medium appear to be reflexive just because both are not separate.

Separation and reflexivity are appearances both for the subject and the object, and it is useless to adhere to one part trying to deny the other, as science has attempted. Intelligence and being coincide —at least from the point of view of intelligence, since this one is incapable of perceiving itself. This reflexivity, this intelligibility, is the plane of universal synthesis itself. But this has also a physical translation. The evolution of a vortex in six dimensions in Venis coordinates could be a good example of the intersection of a naturalistic view with the transcendental plane.

These ideas can be applied to both the mediate and immediate knowledge of Nature. From the operational and formal point of view, all observable knowledge of physics can be included in the principle of dynamic equilibrium. But from the point of view of immediate cognition, one can hardly sustain oneself in the contemplation of the instant without inertia —to such an extent that ball rolling in the void has captured our subjective idea of succession. However, these principles, which seem now much more demanding from the intuitive point of view, since they are much more full of content, are not based on the separation of nature and therefore make their forced "unification" by man less necessary.

Truly, that transcendental plane is the one in which transcendence with respect to Nature and its return to it coincide —but in truth the only nature that is transcended here is that relative to the inertia of habits, what we call our "second nature". Here we could say with Raymond Abellio: "The perception of relationships belongs to the mode of vision of the "empirical" consciousness, while the perception of proportions is part of the mode of vision of the "transcendental" consciousness" [64].

But, of course, in modern science there are hardly any proportions, because all the units we handle are a heterogeneous and unintelligible jumble of quantities —that's why we rely more and more on computers and their programs for data crunching.

The term "transcendental" can only have some meaning for those who deal with the intelligibility of knowledge, not for those who are simply content with its formalization to obtain predictions. However, here we are making it descend to the very core of physical principles and of that eternal unknown we call causality; and this can be done both in a qualitative and a quantitative way.

No physicist or mathematician needs to prove the existence of the complex plane or the complex manifolds, however advisable it may be to review its foundations; much less could we prove the existence of a transcendental plane of synthesis, since the word "transcendental" means that it is the condition of knowledge. Motion is shown by walking, and knowledge by understanding.

Moreover, to speak of the "existence" of such a plane in relation to the world of objects and measurements is not only out of place but completely reverses the situation. Santayana said that essences are "the only thing that people see and the last thing they notice"; from a perspective in line with what we have already said, and which of course has little to do with the usual narratives and cosmologies, the entire experience is a transition between an unknowable but measurable matter and a space that is diaphanous to knowledge but immeasurable. Existence itself is that process of awakening, but with dissimilar rhythms for all kinds of systems and entities.

This our condition "between Heaven and Earth", between the diaphanous and the measurable, is not a mere philosophical or poetic remark, but it determines the whole range of our possibilities of knowledge, which began with the acts of counting and measuring, of arithmetic and geometry, and which have been successively developing with calculus, algebra and everything else until we got here. And we can not only see that it determines it, but also appreciate that there is always a double current, a double movement, descending and ascending.

Nor is it to be believed that this plane of essences is reduced to the mathematical aspect; on the contrary, this is only one possible expression of an infinity of modalities. In our time we have come to believe that complexity exists only in numbers, computers and models, but the richness of phenomena has always been infinite, regardless of any quantification. Our perceptions, as our thoughts, are also a fleeting part of the absolute.

\*

Technoscience is a continuum of practice and theory that dictates what seems acceptable to both. On the other hand, it is not ideas that determine our actions —it is what we do and what we want to do what determine our ideas.

For today's technoscience, to see intelligence as something totally separate from Nature is an indispensable condition for recombining any aspect of nature at will: atoms, machines, biological molecules and genes, and all the possible interfaces between them under the least restrictive criterion of information.

But granting this convenient separation of domains to manipulate them with the least possible restrictions, paradoxically entails unnecessarily restrictive principles, as is already evident in fundamental physics, which is also the founding ground of our overall commerce with Nature.

The reintegration of intelligence into the unity of being is absolutely contrary to the liberal principle of prior separation in order to recombine without restrictions. The mere possibility of an intelligence in Nature threatens to short-circuit this separating intelligence who has established herself as the ultimate arbiter.

And yet, as we have seen, the idea that there is feedback in the orbit of a planet or an electron is less contradictory than the usual picture that we are dealing with a cannonball trapped in a field. In fact, it is not contradictory at all: it is just absolutely disconcerting, as well as inconvenient, for most of us.

We want to see on the one hand a separate intelligence, and on the other a completely inert matter, and between these two fictions, the evidence of a perfectly impersonal, pre-individual consciousness without qualities becomes totally inconceivable.

There is nothing extraordinary about the fact that apparently heterogeneous bodies seek to attain the homogeneous condition with the environment from which they have emerged, and that they cannot do it without both inner and outer action. The particular aspect of this balance is indiscernible from the intelligence of that entity or system, which cannot but participate in the universal intelligence. Were it not for our involvement in this last one, our very intelligence would only exists subjectively for ourselves.

It is not strange at all, but it is totally inconvenient for the practices in which we are immersed, for the horizontal, indiscriminate remixing that aspires to dissolve all natural boundaries.

*

In the immediate postwar period, around 1948, under the shadow of the Manhattan project and other military programs, three new "theories" emerged that consolidated the new "algorithmic" style of the sciences: quantum electrodynamics with its endless loops of calculations, information theory, and cybernetics or modern control theory. To this was added, five years later, the identification of the DNA helix, which would soon be reduced equally to the category of information.

Except for the calculation feats that are its only justification, the quantization of the electromagnetic field was on a theoretical level a superfluous undertaking that added nothing new to the known equations. The funny thing is that its promoters had to be averting terms like "self-interaction" and "self-energy" constantly popping up in their faces. Terms like this seem to sound undesirable for such a clean and fundamental theory as QED, and moreover in physics it is assumed that any kind of feedback only can result in stronger non-linear effects.

Thus, while fundamental physics fought tooth and nail to conjure up the idea of feedback, cybernetics had to assume that feedback is a weak, emergent property of highly organized systems made up of "fundamental" blind blocks. At the same time, information theory hacked Boltzmann's mechanical-statistical entropy into a new brand, rather than the irreversible entropy of thermodynamics. To question the foundations of the past theories would have been out of place, so theorists were content to generalize heuristic procedures, and it could not be otherwise since this science-technology continuum does not demand anything else.

We have seen that our very idea of celestial mechanics, of calculus, and of the mechanical-statistical interpretation of the Second Law are based on flagrant rationalizations —not to speak of more recent developments. Even the explanation of the functioning of our heart is based on a rationalization that tries to ignore the monumental evidence of the role of the breath in the global dynamics of blood circulation.

The only reason we maintain these mirages is a very powerful one, because they reaffirm our idea that we have a separate intelligence, while justifying our indiscriminate intervention in Nature, a Nature that we very conveniently want to reduce to blind laws and random processes. Men of science find it shocking, to say the least, to talk about the transcendental in knowledge, but all modern scientific knowledge is based on an ego pretending to be transcendental that in the end became trivial.

[147]

## 16. The religion of prediction and the knowledge of the slave

In calculus, infinitesimal quantities are an idealization, and the concept of limit, provided to support the results obtained, is a rationalization. This dynamics going from idealization to rationalization is inherent to the liberal-materialism or material liberalism of modern science. Idealization is necessary for conquest and expansion; rationalization, to colonize and consolidate all that conquered. The first reduces in the name of the subject, which is *always more than* any object x, and the second reduces in the name of the object, which becomes *nothing more than* x.

But going to the extremes does not grant at all that we have captured what is in between, which in the case of calculus is the constant differential 1. To perceive what does not change in the midst of change, that is the great merit of Mathis' argument; that argument recognizes at the core of the concept of function that which is beyond functionalism, since physics has assumed to such an extent that it is based on the analysis of change, that it does not even seem to consider what this refers to.

Think about the problem of knowing where to run to catch fly balls—evaluating a three-dimensional parabola in real time. It is an ordinary skill that even recreational baseball players perform without knowing how they do it, but its imitation by machines triggers the whole usual arsenal of calculus, representations, and algorithms. However, McBeath et al. more than convincingly demonstrated in 1995 that what outfielders do is to move in such a way that the ball remains in a constant visual relation —at a constant relative angle of motion- instead of making complicated time estimates of acceleration as the heuristic model based on calculus intended [65]. Can there be any doubt about this? If the runner makes the correct move, it is precisely because he does not even consider anything like the graph of a parabola. Mathis' method is equivalent to put this in numbers.

How much can we synthesize knowledge? If there is no way to prove that an algorithm is the shorter or the faster, there can be no explicit limit to its comprehensibility either, and the same applies to all formal knowledge. However, there is an informal but effective guide both to synthesize knowledge and to improve its quality: always pay attention to the invariable mean, to what does not change in the midst of change. And although it is an informal principle, it is always possible to recognize its lineaments: the example above is clear enough both in the real world and in formal analysis.

In fact, it should be said that it gives us an incomparably straighter and simpler path than the "alternative models", in this case the established methods of standard calculus and the analysis of algorithms for tasks in artificial intelligence. Over time we can find an infinite number of instances with a potential for convergence at least as great as the potential for divergence of the usual analysis in terms of change.

Examples of idealizations are the inertia principle, infinitesimal quantities, point particles and the existence of single individual systems in quantum mechanics, reversibility at the fundamental level, the global synchronizer in both Newtonian and relativistic physics, universal physical constants with dimensions that are isolated from the environment, or space as a differentiable manifold. These are force-ideas born out of an intimate need of a subject and a culture, expressing the will to expand till the maximum extent.

This idealization required an intense faith in the program that today, when even the rationalization phase seems unnecessary, we can only underestimate.

It is for the present moment to see through idealizations and rationalizations, not to try to move beyond them, since they are already the expression of extremes and of an extreme exercise of abstraction and experimentation. That being reified is only a thought, but the thinker is a thought as well, and the real thinking process, the Logos pervading the world, can only reverberate in a distorted way between both.

The idea of an invariable mean and the idea of dynamic equilibrium, principles so far removed from the scene by long overdue idealizations, are the two sides of the coin and define a new spectrum of relations between formal and non-formal knowledge, between duality and non-duality.

To think that today we are closer to a "theory of everything" in physics than we were in Newton's time is just a mirage. We could increase the amount of knowledge by a trillion without getting any closer to any ultimate truth. In practice, exhaustion comes much sooner than true knowledge; but just as in aging, claudication results from the inability to eliminate and renew oneself.

Nor are we any closer to "unraveling the mystery of consciousness" than in the time of Leibniz and Newton. In fact we are closer to that time than to "the final solution"... among other things, and to begin with, for failing to recognize things like why the calculus works. We haven't even suspected that both things may be connected.

Formal knowledge can increase indefinitely without ever reaching what we are experiencing right now, and the advance in informal knowledge only takes place by the direct recognition and appreciation of something always present but unnoticed. Just as power and knowledge limit each other, so do formal and formless knowledge, but there can be no law specifying their relationship.

So there is no "transcendental horizon" of gradual approach to truth for accumulated social knowledge, but neither for knowledge prior to formalization. For the same reason, in formalized knowledge the possibility exists of "transcending" forms, at least in the sense of leaving behind idealizations and rationalizations.

Much more could be learned today by gently modulating the states of particles than crashing them in accelerators, just as more could be learned by developing a proper theory of the extended particle than speculating about the origin of the universe expanding from another point. But to do this we would have to remove the extreme weights of our bets, that is, the unnecessary restrictions of our present theories. To say that in present day science "theory is superseded by correlation" would be the clearest example of extreme propaganda for the prevailing theory and practice. In fact, if many laboratory experiments do not have a greater significance today, it is because a standard interpretation of facts is immediately enforced even if the theory has nothing to do with it.

It is known how in 1956 Bohr and von Neumann came to Columbia to tell Charles Townes that his idea of a laser, which required the perfect phase alignment of a great number of light waves, was impossible because it violated Heisenberg's inviolable uncertainty principle. The rest is history, but Bohr's and von Neumann's words are not recorded. This is no exception, but what happens constantly and routinely. It is clear that a theory that subtracts infinity from infinity every time is needed can predict anything, especially after the events. They call it "powerful methods".

So, there is no observation that quantum electrodynamics can not rationalize. And the same goes for the two theories of relativity, statistical mechanics, classical mechanics or calculus. This is what means to speak of "powerful methods", what puts the search for truth in a desperate situation. Ptolemy's epicycles were also undoubtedly a powerful method, as they could cope with all kinds of celestial observations with an unbeatable predictive power for the time.

The problem with formalized knowledge is that once we accept a standard and push all sorts of things in, it is terribly difficult to get out of it. In the case of relativity, a certain reform of classical mechanics was accepted because, in addition to the urgency of solving blatant contradictions, the unification with electrodynamics promised an even greater expansion. Another issue is that since Gauss and Weber everything could have been done differently, sacrificing other things and obtaining different advantages in return.

Even if the amount of accumulated knowledge does not increase one iota, the formalization of knowledge goes through mathematics, and mathematics is always capable of expressing any concept and relation of concepts in a new and even unrecognizable way. We have already seen this: the same phenomena can be described by saying that the movement of bodies is determined by external forces, as by saying that the bodies themselves determine their movement. If this does not surprise you, nothing will do.

It would be much more attentive to reality to try to recombine to some extent our ideas than to try to recombine the whole world to its smallest parts only for the sake of manipulating them. It would also be more attentive to the quality of our knowledge, too. In formal knowledge anything can become anything else in the end with due transformations, but here we are not talking about arbitrary transformations, but about historically possible transformations, transformations of meaning and sense.

In other words, one cannot change everything at once, far from it. But there are clear lines of action to turn a glove inside out without breaking it, and here we have been talking about such lines. The same quantitative increase in knowledge has nothing to do with improving its quality, and now is rather the symptom of a great imbalance.

Conversely, the improvement of quality is intimately but not expressly related to the balance between formal knowledge, based on change, and the awareness of the invariable and undivided, which does not seek infinity because it knows it has it within. This balance also depends on the harmony between principles, means and ends; on how the circle of interpretation closes on the principle using the straightest means.

Even if mathematics does not care about what reality is, it still depends on the form, which thus becomes a reality on its own. It is through mathematical physics that it has come out of her splendid isolation, but physics has used and abused mathematics to conquer the world rather than to see it. Of course, the opposite is always equally possible: math can use physics to investigate its own relation with reality, in a way totally different from the one used up to now.

Since reality is first and foremost what has no form and what is the support of forms, but mathematics can begin to glimpse another relationship with it through the recognition of unity in the invariance of change. It is in this sense, as an interpreter of physics in its most basic sense, that it can transcend itself and access the transcendental plane.

*

In the Platonic dialogue *Meno*, Socrates poses questions to a young slave with no more culture than his knowledge of Greek by drawing a square on the floor and then another. After a skillful interrogation, asking him about the length that the side of the second square must have in order to double the area of the first, and after intermittent phases of stupefaction, he manages to awake in the young man the idea of irrational numbers, the first "great revolution" of ancient mathematics. Socrates prides himself on not having teaching the slave, but of having helped him collect the knowledge just questioning his previous answers.

As always, various readings can be made of this famous pedagogical moment, and Gómez Pin, in a magnificent book whose subtitle is precisely *The Knowledge of the Slave*, takes the reasoning further until the slave finds out himself the idea underlying infinitesimal calculus [66]. Certainly, for us it seems that one can pass in two hours from irrational and real numbers to the inception of calculus. But why then did it take more than two thousand years since Plato's time? Many great minds devoted to this problem not

[154]

only two hours, but a great part of their lives, without approaching the crux of the matter.

The culture and knowledge we treasure as a society serves both to inscribe new things in our minds and to erase them. But high-level knowledge, very specialized wisdom, is a very delicate plant. Mathematicians naturally tend to think that theirs, not that of philosophers, is the true factory of ideas; to this it could be argued that pure mathematicians handle far more rarefied concepts, and that they only have delivered true ideas when they are philosophers to begin with, as in the case of Pythagoras, Descartes, or Leibniz, or at any rate "natural philosophers" and physicists like Newton.

The advanced concepts of mathematics only serve at the cutting edge of instrumentalization. What can a modern-day physicist or mathematician convey even to an educated public about his research? Gómez Pin says that the categorical knowledge inscribed in language is more basic than mathematical knowledge: differences such as quantity and quality, or the category of measure, which is precisely a mediation or synthesis of both.

No doubt there is quite a lot of truth in this; there are ideas more basic than mathematics, which are not exclusive to any discipline, and which channel the drift of mathematical concepts. These ideas are at the basis of a cultural syntheses of an entire epoch or civilization. And finally, there are symbols, which do not even pass through the antinomies of ideas and concepts but can adopt them for mere external convenience. In another time, these symbols were supports of various traditions that sought to transmit a knowledge beyond the reach of forms.

There is a formless knowledge that is truly our common root and soil; the ideas are like the sap that ascends through the trunk of the diverse cultures, and the mathematical concepts, would be rather at the level of the branches, the leaves and the fruits. If fresh sap ceases to ascend, what we have is the autumn of a culture, the season of dry leaves. And instinct is also a part of the implicit knowledge, but shaped and conditioned by the social environment. No one tells us how to catch fly balls, but whether one does it playing baseball or practicing different skills depends on the culture and environment.

The point is that instinct is presented to us as opposed to reason when it is obviously reason that is opposed to instinct, to nature trapped within us. Or perhaps it should be said that it is not even reason, but a series of rationalizations. It always seemed to me that Newton's explanation of the ellipse contradicted not only reason, nor reason and intuition, but reason, intuition and instinct; and it also seemed to me that if people accepted this sort of explanation, it was certainly not because of instinct or reason, but because of a certain interest. The same could be said of wanting to enclose the universe in the laws of mechanics.

Let us see what reason and instinct tell us about the next question. Following the same logic of his amendment of standard calculus, Miles Mathis states that the true value of the constant $\pi$ in any situation involving motion is not 3.14... but 4. This is something that he has argued in detail showing, to begin with, that he has read Newton and other classics much more carefully than his critics [67].

Mathis is simply saying that length is not the same as distance and that to advance around a curve like a circle one has to move in two directions at once. It is just adding up the vectors, not taking the graph literally. Of course *pi* gives us 3.14... in relation to the diameter in a straight line as if we were measuring it with a string; but the point is that moving around a curve involves simultaneously a velocity and an acceleration.

If anyone is in doubt, he can ask himself whether he will use the same fuel driving in a car for 3.14 kilometres in a straight line than driving around a circle 1 kilometre in diameter. And if the difference is attributed to friction, he may wonder if an object in space with an external impressed speed, which according to the principle of inertia should continue with the same direction and speed indefinitely, can go round and round indefinitely. As far as we know no law of inertia with perpetual motion has been enunciated for circular motion.

[156]

And yet that is what Newton's fundamental lemmas at the beginning of the *Principia* imply. According to Newton and all celestial mechanics since him, a planet in a perfectly circular orbit could orbit around a central body forever, exactly like a perpetual motion machine. But I think it is clear that friction has nothing to do with this here, since anyone understands immediately that a contribution of force is needed not only to change speed, but also to change direction.

The question, then, is not how it is possible for Mathis to say what he says; the question is how it is possible for Newton and Euler and Lagrange and Laplace and all the others to have been able to accept this for over 300 years without a blink, and how we can continue to accept it without giving it the slightest attention. Mathis himself asks this question repeatedly, and hesitates between the hypothesis that they did not perceive it and the hypothesis that they did perceive it but hid it.

Surely they were aware of it. But we underestimate the ascendancy that over them had certain ideal of science and nature alike, that of a clockwork governed by a clockmaker. Then, simultaneously with the perfectly utilitarian idea of expanding the realm of calculus at any rate, they felt morally justified that this contributed to bringing us closer to an ideal of nature perfectly passive in relation to its creator. This intimate mixture of utilitarianism and disguised theology, in which a separate personal creator and an equally separate cognitive subject melt together, still has a great ascendance even today; but the important thing for the instinct is not to know itself. On the other hand, while eager to play the game, surely they believed they were taking "the shortest path" to their goal —which was also their ideal of efficiency. Besides, no one could imagine that calculus understood as reverse engineering — truth could not escape- was amenable to such kind of mistakes.

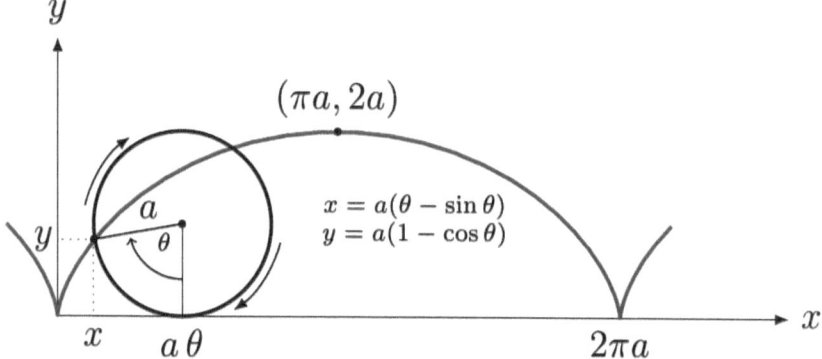

Another way to visualize this issue, if necessary, is by means of the curve known as cycloid, drawn by the rim of a wheel rolling in a straight line without slipping. The length of this curve is exactly 4 diameters or 8 radii, and the motion it describes is extraordinarily elusive only if we are thinking of the ordinary circular graph.

It is ironic that the first diligent study of this curve was made by Galileo, the father of the principle of inertia, —a principle so elusive for the time, that it took fifty years to consolidate and only after the contributions of Descartes and Newton-, in order to calculate quadratures. As he made his laborious calculations, he was pushing with his own hands time and again the best possible counterexample to that idea of the rolling ball that had already taken possession of his mind.

As is well known, Galileo opposed Kepler's elliptical orbits and and continued to adhere the ideal of circular orbits. Huygens made an even more thorough study of the cycloid to improve the accuracy of the pendulum clock, whose very idea was due to Galileo. Later Newton had the matter at hand again while dealing with the celebrated problem of the brachistochrone curve, whose solution is precisely the cycloid.

I don't think that the subject $\pi = 4$ deserves more explanations; and whoever needs more information already knows where to find it. If we want to express it in terms of limits, it can be said that bodies travel in the curves to the limit of the shorter legs of the triangle, instead of the limit of the longer leg. It is called the "Manhattan metric", but it casts by the way a deep shadows on the basis of the vast majority of metrics. However the purely "technical" question pales in comparison with the effect which it should have on our minds to detect such holes going back to three or four centuries. This alone should radically change our perception and relationship with science.

What the derivations of Huygens and Newton precisely show, round about the crystallization of calculus, is the delicate pass from talking about proportionality in forces to talking about their equality. As in the case of the definition of the central forces, Newton is extremely cautious and avoid expressly to speak of equal forces; but once calculus paved the way, the equality was taken for granted. With the emergence of the "vanishing quantities" the constant proportions of geometrical figures left the scene forever.

It is possible that a great deal of the mismatch between the calculus and the continuous proportion is due to the portion of motion vanished in the representation of graphs, of which the kinematic circumference is the best example. It has been said that the continuous proportion belongs to the realm of statics, in contrast to the changing world of calculus —but in reality the opposite is true, it is calculus, the tool designed to describe change, that adheres to the static figures. This should have implications for a modern theory of proportionality, a possibility that was uprooted and wiped off the map by the same operation of standard analysis.

Calculus, as it has been understood, really involves a inversion in our natural view of the physical world. As Krishna Vijaya observes, instead of determining geometry from physical considerations, deriving from it the differential equation, since Leibniz and Newton the differential equation is set up first and then one tries to solve it to get the physical answers. Both methods are far from equivalent, but the same belief in the reality of the differentials follows from the procedure adopted. It is still necessary to reverse this method in order to open the eyes and recover the right perspective [68].

Ideally, description and prediction should be balanced, as should memory and anticipation, with which we constantly and reflectively create our perception of time. Inevitably, if there is a perception of an "unreasonable effectiveness of mathematics" in predicting natural occurrences, it is because a great part of the description has been obscured or concealed. This creates enormous cognitive dissonance in both our perception of nature and our idea of science and knowledge. Modern science is always urged to be more and more "creative" in order to be less and less aware of the nature of its manipulations. But a way exists in which the liberation of Nature and the liberation of the knower coincide.

One can understand Nature without calculus, as one can predict things of Nature without the least understanding of them. Venis' morphology is an instance of the first, and modern physics is certainly the best possible example of the second; but that does not mean that one kind of knowledge excludes the other; everything depends on how the connections are elaborated and derived. The crux of the issue is that modern technoscience is far more interested in manipulation than in understanding, and to such an extent that understanding becomes inconvenient. In turn this limits the degree of disorder that we humans can create.

Philosophers have repeatedly complained that calculus involves a geometrization of motion, but they have never been able to substantiate their claims. Now that they have it, they can seize the opportunity to delve further into the subject.

The cycloid and the wheel could serve as an indication of a different quadrature than the one Galileo intended; rather the opposite, though still with an important point of contact. It is well known that throughout history, and for very different cultures, the circle and the square were the symbol of Heaven and Earth, the active and the passive, the dynamic and the static.

However, the understanding of what is "dynamic" and what is "static" was definitely reversed in the short time span from Galileo to Descartes. Before this reversal, motion and changes in extension could reflect change, but rather accidentally; and so potentiality and actuality in Aristotle had an incomparably broader meaning than that now attributed in physics to potential and kinetic energy —a pair already defined in terms of motion.

Physics takes off when it starts to define everything in terms of extension and motion, although in a very unclear and indistinct form, one is aware that not all physical reality can be reduced to motion and extension.

"What moves does not change and what changes does not move." The closer we get to our time, the more important motion becomes, and the further we move away, the more secondary and related to appearances it is considered. But in this continuous transformation there is permanently a double movement; for it is evident that calculus has also frozen motion and not only reduced it to geometric forms but has assumed its static quantitative relations.

This double movement of ascent and descent, of condensation and volatilization is really a natural and spontaneous process that can take place at different levels —individual, collective, physical and non-physical. The history and cycles of different civilizations can also respond to this double pattern, and the same history of science shows intermittent evidence of it.

FIG. 6

Rivers of ink have flowed about the Cartesian dualism that separates mind and extension, but much less has been said about the duality inherent in physical quantities, in which an extensive part always coexists with a non-extensive one, being now mediated by motion itself. From this stems pairs such as space and time, force and mass, vector and scalar, intensive and extensive quantities, until the extremely complex measurement units of today.

This duality of physical magnitudes, antinomic or not, is nothing but the resolution in the plane of motion of that first duality; and there is no reason why this endless dialectic should have an end, since it is the very deployment of reason. Only ignoring our role in all this could we believe that the cause of consciousness can be found at some explicit level of causation. The thinking substance is already threaded into the description of any level of physical causation since the times of Galileo and Descartes and even much earlier; so in vain do we wait for the paradoxes of quantum mechanics to solve these enigmas for us.

There is a big difference between an idea, and even a mathematical idea, and the manipulation of mathematical symbols. Profound ideas arise when concepts try to return to interpretation or representation, and when representation pays due tribute to the unrepresented, to the implicit in knowledge. But the current drift of science prevents the full circle of this motion that should always be expectant towards the most basic, the really fundamental.

The world would only be an illusion if it could be reduced to extension and motion, yet science fails to tell us what it can be beyond these attributes. Here is its limit, but it would have to be a fertile limit. And the limit between that which cannot be represented and that which is representable passes precisely through this double movement: this is the best exponent of true activity, in Nature and in Spirit alike. We can find exponents of this double movement on many levels, from the electron to the constitutive relations of materials, from our own breathing to the movement of the vortices in the Venis sequence, from the process of individuation to the same self-consciousness.

The swastika offers us a good example of a symbol containing implicit knowledge. It is first of all a sign of the Pole and its action on all things in the world; but it can also be the clearest expression of squaring the circle through motion, of the double movement itself, of the dynamic equilibrium, or of the reciprocity intrinsic to the Law. Surely there was little concern with the issue of calculus in prehistoric times, and yet we always can actualize and give voice to any mute expression of reality.

*

Change itself is only the visible side of the transcendental plane, but this change has an infinite number of aspects that escape our limited ideas on motion. In this sense, all modern physics and science in general continue to have a huge Platonic deadweight, and we have gone from metaphysics to mathephysics without hardly noticing it.

The mere fact of thinking that there are fixed physical constants or identical particles is already a total lack of taste when it comes to describing Nature, an eloquent exponent of our incapacity. And yet, in a framework that aspires to "maximum predictive power", these assumptions are indispensable. Can identical electrons exist, as if they had just come out of a factory, independently of our complex web of assumptions, constants, and measurement conditions? And yet within this framework it is nearly impossible that it could be otherwise.

An electron, like ourselves or any other entity, can only be an ephemeral configuration that changes from moment to moment. It is not the individual what matters, but the process of individuation, which constantly connects what appears to us as the part and the whole. This connection is pure activity, balance or fight at present, not a frozen event that creeps through the universe for thirteen billion years.

The analysis seems to dissolve everything, yet the fluidity of the world escapes it completely. How can this be possible? That's an excellent question that should be answered. Because, currently, it is not even approaching the real flow of things and the great unity of life, but on the contrary is moving further and further away from them.

[163]

In fact this is a fatal question for modern science. Because it is clear that the immutable cannot be the object of knowledge, and that if something can be known, it is a matter of change, and nothing else. How is it that analysis, which is the study of change, ignores almost everything about it? Is not it amazing that scientists do not ask themselves more questions about this? But there is not even a proper frame to pose the question.

What can be predicted is an expression of regularity, and in that sense, of law. However, to subordinate everything to prediction one is forced to separate and isolate aspects of the processes to cut out the forefront from the background and the context. But even if the ultimate background could be neutral, the context never is. Any context already flows in a definite direction, but now we have a different direction, because to be guided by predictions alone is like running with blinkers.

Modern science tries to compensate for the void of analysis and prediction with disciplines of a "synthetic" nature such as cosmology and the theory of evolution in an attempt to reconstruct the contexts and directions that analysis has destroyed —but in doing so completely relies on the assumptions established by the analytical criterion. And so the whole of cosmology depends implicitly on the principle of inertia, and the theory of the evolution of life on inertia plus the purely random nature of the processes. We have already seen that the first assumption is unnecessary and contradictory, and the second false.

Since the analytical-predictive part took precedence, and the synthetic part relies on its assumptions, the whole is completely unbalanced. The underlying idea is that we can dissolve everything, and then put it back together again and give it a direction and a narrative. But in this way it is impossible to respect the reality of things.

The "analytical part" already has a direction, and therefore cannot be separated from the synthetic part. In other words, nature already has its own narrative all the time, so it does not need any other one as a supplement.

[164]

Venis' infinite sequence presents us the unity of the analytical and morphological aspects, of balance and direction in Nature. This unity is the real challenge of modern science and of science in any age; the contrast between its perspective and the perspective of modern analysis and synthesis will prove to be supremely instructive.

Heraclitus understood the world better than contemporary scientists; if he already disdained Pythagoras' knowledge as polymathy, as a mere variety of learning, it is better not to imagine what he would think of our theories and specialities. The problem is not the quantity of knowledge, but its quality; and mathematics can be a master of deception in this regard. You end up having the knowledge you want, but the point is what kind of knowledge you want.

*

One thing is the spontaneous organization in Nature and another is the recurrent presence in it of a certain mathematical constant. It has yet to be demonstrated that there is any kind of functional link between both, and if there is, nothing can yet be said about its relevance. What admits few doubts is that fundamental physics can be interpreted most directly as a phenomenon of self-organization, through relational mechanics and the retarded potentials; as well as with the due recovery of the "thermomechanical" irreversibility, in fact the only real dynamics.

But theoretical understanding alone is incapable at this point of modifying in the slightest the highly dissolving and destructive drift of current science, so in tune with the rest of the social dynamics —we cannot change ideas without first changing what we do and what we want to do. Today man does not seem to be between Earth and Heaven, but between a Nature in perpetual setback that we only see as a resource pool and machines that materialize the human spirit to exploit both the external nature and our internal nature.

We say "human spirit" because not only our intelligence but also our will is materialized in the machine: and the substantial unity of both has become inconceivable to us precisely because of the growing separation made possible by machines. Yes, machines, like other human creations, are an obvious crystallization of a certain water and fire, a feminine desire and a masculine will: a small abbreviation of nature isolated from the rest that tries to perpetuate its movements at the expense of the environment. We pay dearly for this simulacrum of closure that is not closed at all nor can it be; in fact, perpetual motion and its *telos* never ceased to exert a fatal spell on us.

Throughout this writing we have been indicating a common axis that runs through man, nature and the machine, and must undoubtedly go beyond them, since we know of them nothing but momentary planes of manifestation. If current technoscience separates in order to recombine at will and unleash the universal confusion of planes, we may well do the opposite: see where different planes coincide in order to survey their vertical, their height and depth.

# 17. Information theory and the zeta function

It has been said that if the Riemann hypothesis were solved, all the keys to cryptography and cybersecurity could be broken; no one has specified how that could lead to faster factorization methods, but at least it reminds us of the close relationship between a hitherto intractable problem, cryptography and information theory.

Such speculations are only based on the fact that the Riemann zeta function establishes a connection between prime numbers and the zeros of an infinitely differentiable function that provides the most powerful method for exploring this field, prime numbers being the basis of classical cryptography. But we have already seen that the zeta is way more than this.

Today the technological revolution is synonymous with the digital revolution and with a category, information, which seems to pervade everything. All previous developments in science and technology converge in it and pass through it. Even physicists now think of the universe as a gigantic computer; there was much discussion about whether the world was made of atoms or stories but in the end it was decided that it was made of bits and case closed.

What is important about information theory is not so much its definitions as the direction it imposes on everything; changing that direction is equivalent to changing the direction of technology as a whole. The unequivocal direction, which obviously inherits from statistical mechanics, is that of decomposing everything into minimal elements that can then be recomposed at will.

For statistical mechanics there is no direction in time: if we do not see a shattered vase recompose itself and return to the table, it is only because we do not live long enough; if the pieces of a dismembered corpse do not get back together and walk again as if nothing had happened, it is only because we are not in a position to wait $10^{1.000.000}$ years or something similar.

Information theory is not concerned at all with the reality of the physical world, but with the probability in its constituent elements, or rather the probability *within its accounting of the constituent elements*. On the contrary, the physical world is just a pool of resources for the sphere of computation, aimed to be independent of the former.

Is this statistical view a neutral position or is it simply a pretext to manipulate everything unencumbered? There is no single answer to this. Statistical mechanics and information theory are successfully applied in countless cases, and in countless cases they could not be more irrelevant. What is worrying is the overall trend they create.

It is my conviction that a shattered corpse is not going to recompose itself, no matter how fabulous the period of time we choose. And this is not rhetoric; rather, large numbers are the rhetoric of probability; an inflated and poor rhetoric, as it ignores the interdependence of things.

This interdependence, this infinite network of interrelations, is what makes things what they are. Statistical mechanics, and its daughter information theory, are the most general framework to deal with random independent elements; the Riemann zeta function, the most direct and elegant way of encompassing an infinite series of elements, the numbers themselves, which are both independent and dependent, apparently random and simultaneously containing an infinite net of relationships.

Then it is only a matter of time before the information theory and the zeta function meet together. Today numbers no longer seem to exist to understand the world, but to crush and squeeze it, and with it all of us. Information theory have become the funnel, the fearsome event horizon for everything; but in turn the zeta function could become the event horizon for the Information Age itself as the very concept of information, so highly generic, is refined and given content.

However, somewhat surprisingly, there is virtually no literature on an explicit relationship between the zeta function and information theory. In our search we only found a brief note due to K.K. Nambiar, in which he showed a connection between the capacity of the channel and the function, with a "Shannon series" and a "Shannon zeta function". The idea was worthy of much further elaboration, discussion and development. Besides, the equivalence between the zeta function and the classical sampling theorem fundamental in signal processing seems proved [69].

The same author issued an even shorter note establishing an electricity equivalent of the Riemann hypothesis in terms of the power dissipated in an electrical network; of course more elaborate equivalents in terms of electrical potential have also been established, but here we are more focused on thermodynamics. Infinitely many waveforms can be created with radiation patterns containing the zeros, the point is how to modulate them. Already in 1947 Van der Pol had created an analogical electromechanical device for computing zeros from the zeta [70].

If not directly with information theory, at least there are many more works connecting the function to closely related aspects: entropy, statistics and probability [71]. Here we will focus on the concept of entropy.

Entropy and information become almost synonymous, and if in thermodynamics entropy was seen as the loss of usable energy, it will now be seen as information loss. Shannon's information was initially called *surprisal* or *self-information* —a message should contain something new.

The confusion of entropy with disorder is due to Boltzmann's rationalization in attempting to derive macroscopic irreversibility from mechanical reversibility, quite a tour de force by itself. It was Boltzmann's use of the concept of "order" what introduced a subjective element. Clausius original energetic conception of entropy, stating that the entropy of the world tends to the maximum, was, if nothing else, far more natural.

Order is a subjective concept. But to say world is to say order too, then also the idea of world is inevitably subjective. However subjective they may be, "order" and "world" are not just concepts, they are vital aspects of every organized entity or "open system". Yet even our very intuition of order and entropy are entangled and confused.

Not that entropy increases disorder. On the contrary, the tendency towards maximum entropy is conducive to order, or better said, order has more potential for dissipation. As Rod Swenson put it: *"The world is in the order production business, including the business of producing living things and their perception and action capacities, because order produces entropy faster than disorder"*[72].

Information theory is a formal and objective framework but it cannot escape from the reflexive turn in the representation and use of data. Information as an object is one thing, but the information as an environment with which an open system interacts and helps to shape is another. Although in black box mode, in many cases the interpretation is already part of the system's behavior.

The ambiguities and limitations of information theory have given way to more inclusive frameworks, such as Luciano Floridi's philosophy of information, which, to put it very simply, considers semantic information as data + questions [73]. Floridi's approach still remains a clear heir to Cartesian dualism and idealism, and in this decisive division we prefer to speak of two types, external and internal, of information and entropy: there is a kind of information-entropy as a formal object and there is an information-entropy environment including the open systems that interact with it. Boltzmann's entropy and Shannon's entropy are of the first type,

Clausius' entropy and the one that engineers deal with in concrete physical applications is closer to the second.

The connection between entropy and the zeta function is quite recent and only in the 21st century is beginning to take shape. At a very basic level, one can study the entropy of the sequence of zeros: a high entropy would lead to a low structure, and vice versa. It is observed that structure is high and entropy low, and even low level neural networks were already successful in their prediction [74]. Also, of course, we can study the entropy of the primes in the ordered sequence of numbers and their correlation with the zeros; and from there on we already have an unlimited spectrum of correlations.

There are other non-extensive types of entropy independent of the amount of material, such as Tsallis entropy, which can be applied to the zeta; these types of entropy are generally associated with power laws, more than exponential laws. But as in thermodynamics in general, the definition of entropy can vary from author to author and from application to application.

In order to unify and generalize so many definitions of entropy, Piergiulio Tempesta has proposed a group entropy [75]. Whether in physical contexts or in complex systems, in economics, biology or social sciences, the relations between different subsystems depends crucially on how we define the *correlation*. This is also the central issue in data mining, networks and artificial intelligence.

Each correlation law may have its own entropy and statistics, rather than the other way around. Thus, it is not necessary to postulate entropy, but its functional emerges from the kind of interactions that one wants to consider. Each suitable universal class of group entropy is associated with a type of zeta function. The Riemann zeta function would be associated with Tsallis entropy, which contains classical entropy as a particular case. But these correlation laws are based on independent elements, while irreversibility at the fundamental level seems to reject that assumption. Fundamental irreversibility assumes universal interdependence.

We mentioned earlier Voronin's theorem on the universality of the Riemman zeta function, which shows that any kind of information of any size can be approximated with arbitrary precision within this function —and not just once, but an infinite number of times. In this sense, the entropy of its mapping is infinite. It seems that there is a "zeta code" that could be the object of the algorithmic theory of complexity. But in turn Riemann zeta function, by itself an infinite encyclopedia of correlations and correlation laws, is only the main case of an infinite family of related functions.

One could also find any kind of information in white noise; but white noise lacks any structure, while the zeta function, being well-defined, has an infinite structure. Had Hegel known both cases he would have spoken of false and true infinity; we will see that these Hegelian apperceptions are not totally out of place here.

The *seriousness* of a problem such as the zeta demands that the problem of irreversibility be considered in depth. And considering it in depth means precisely including it at the most fundamental level. Conservative systems are still toy models for this.

The MIT school or Keenan school of thermodynamics, especially with Hatsopoulos, Gyftopoulos and Gian Paolo Beretta, has developed a quantum thermodynamics in stark contrast with the rationalization of entropy by statistical mechanics. The dynamics is irreversible at the most fundamental level. The number of states is incomparably greater than in the standard quantum mechanics, and only a few of them are selected. The selection principle is very similar to that of maximum entropy, although somewhat less restrictive: it is the attraction in the direction with the steepest-entropy-ascent. No great changes are needed in the usual formalisms, what is transfigured is the sense of the whole [76].

The equilibrium formalisms are also retained, but their meaning changes completely. The uniqueness of the stable equilibrium states amounts to one of the deepest conceptual transformations of science in the last decades. The approach is contrary to the idealism of mechanics and much more in line with the daily practice of engineers, for whom entropy is a physical property as real as energy. The theory has many more advantages than sacrifices, and it can be applied to the entire domain of non-equilibrium and to all temporal and spatial scales.

But of course irreversibility can be introduced directly into dynamics as we saw in M. J. Pinheiro's reformulation of classical mechanics, replacing the Lagrangian principle of action with a system with a balance between energy and entropy. This allows the study of entropy generated in classical trajectories under different formalisms and criteria, establishing another general connection between different domains.

In general, for the physicists it does not make sense to change equations that "already work perfectly well". And, in fact, what fundamental physics has always done is to leave aside issues such as friction and dissipation to better stick with the "ideal cases". The principle of inertia is perfectly suited to the task. In this way thermodynamics was carefully segregated as a by-product of ideal physics, which is a curious arrangement indeed.

But we should try to see it the other way around: seeing how the appearance of reversible behavior emerges from a background of irreversibility is the closest thing to finding the magic ring at the heart of nature. Physics today is not about nature, but about certain laws that affect it. Reversibility and irreversibility are like fire and water; only if they intermingle in the right way one can hope to get the real thing.

The ironic twist is that mechanics, with its idealist disposition, has segregated thermodynamics; but in the end it will be information theory, the most idealist offspring of mechanics, that will need to recover everything that has been eliminated if it is to regain its meaning. And the zeta function should play an essential role in that mute and transfiguration of information theory. Information theory should be interested in irreversible mechanics... because it provides more information. And information of critical importance [77].

*

[173]

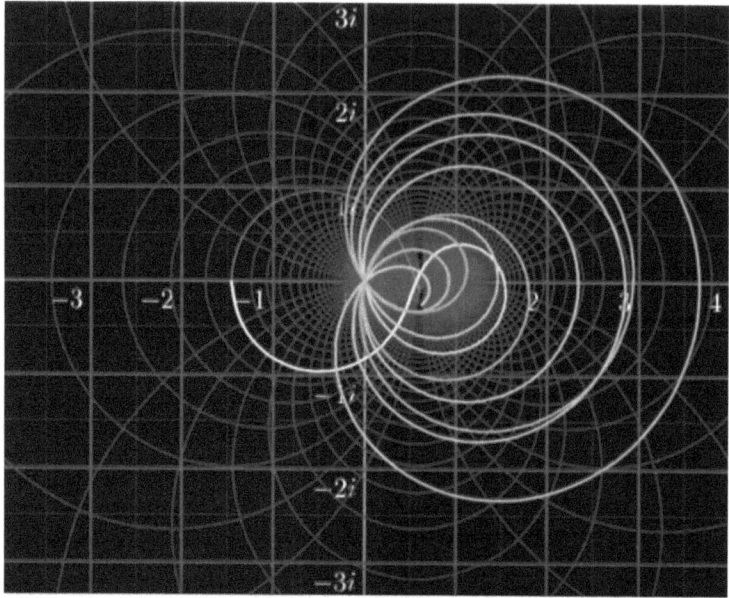

Idealization and rationalization are like the mythical Symplegades rocks that destroyed ships and navigators; only the one who understands their dangers and avoids them will be able to pass to the other side.

We have seen how this alternate process is inherent in the historical evolution of disciplines such as calculus, classical and quantum mechanics, or statistical mechanics. Hardly any experimental evidence can stop the overall trend of rationalization, since rationalization always finds ways to assimilate results and justify inconsistencies.

In philosophy, Hegel gives us the best example of the large-scale sway between idealization and rationalization. In an interesting article, Ian Wright attempts to explain the nature of the Riemann zeta function with Hegel's *Science of Logic*, as the mediation between being and non-being through becoming within the realm of numbers [78]. Old-fashioned as this can sound to many, three things can be said:

[174]

First, that from the point of view of pure arithmetic, this is admissible since if there is a part of mathematics that can be considered a priori, that is arithmetic; while, on the other hand, the very definition of this function affects the totality of integers, and its extension to real and complex numbers.

Second, that calculus is not pure mathematics like arithmetic, although many specialists seem to think so, but mathematics applied to change or becoming. Were it not, there would be no need to compute zeros. The zeta function is a relationship between arithmetic and calculus as close as it is uncertain.

Third, for experimental sciences such as physics the opposition between being and not being does not seem to have systemic implications, and for formal sciences such as information theory this is reduced at most to fluctuations between ones and zeros; however the distinction between reversible and irreversible, open and closed systems, which is at the very heart of the idea of becoming, is decisive and lies at the crossroads of the issue.

In problems such as the one posed by the zeta function in relation to fundamental physics, crucial experiments capable of unraveling the relationship between the reversible and irreversible in systems could soon be conducted —provided that they are expressly sought. For instance, a Japanese team very recently showed that two distinct probes of quantum chaos, noncommutativity and temporal irreversibility, are equivalent for initially localized states. As is well known, this function has been studied a great deal from the point of view of non-commutative operators, in which the product is not always independent of order, so this would bring together two areas as vast as distant up to now [79].

With the rise of quantum computing and quantum thermodynamics, along with all the associated disciplines, there will be no shortage of opportunities to design key experiments. On the other hand, it is becoming increasingly clear that the second quantization, which deals with many-body systems, demands an special spectral approach, as Alain Connes and many other researchers have been emphasizing for many years now [80]. In addition to its theoretical interest, the practical scope of such a spectral theory could be extraordinary —think only of chemistry; but it is our opinion that the segregation of thermodynamic irreversibility prevents us from getting to the bottom of the matter.

It is not that hard either. It is said that electromagnetism is a reversible process, but we have never seen the same light rays returning intact to the bulb. The irreversible is primary, the reversible, important as it is, is just a settling of scores. Even Maxwell equations belong to two different thermodynamic categories.

But physicists have always been fascinated by reversible artifacts, independent of time, which some have seen as the last legacy of metaphysics. And what else? There are no closed, reversible systems in the universe, which are just fictions. Metaphysics is an art of fiction and physics has continued metaphysics with other means with the perfect excuse that it now descends into reality. And no doubt it has, but at what cost?

Previously we gave the example of the outfielder who catches the fly ball as a contrast for both the standard calculus and the now prevalent idea of artificial intelligence. This direct form of calculus, which we all perform without knowing it, could be the best illustration of Mathis' constant differential method. Mathis is the first to admit that he has not been able to apply the principle even in many cases of real analysis, let alone complex analysis, which he does not even deal with. Needless to say, the mathematical community cannot stop to consider such limited methods.

And yet the constant differential is the naked truth of calculus, without idealization or rationalization, and we should not dismiss something so precious even if we do not succeed in seeing how it can be applied. If a constant differential is not found in the tables of values of the function, we can still estimate the dispersion, and whoever says dispersion, may also say dissipation or entropy.

In this way it is possible to obtain an *intrinsic entropy* of the function, from the process of calculus itself, rather than from correlations between different aspects or parts. This would be, at the most strict functional level, the mother of all entropies. It should be possible to apply this criterion to complex analysis, to which Riemann made such fundamental contributions, and from which the theory of the zeta function emerged.

The search for the constant differential may even remind us of the mean-value methods of elementary calculus or of analytical number theory. However the core of this method is not averages, it is on the contrary standard calculus that relies on averages —which without this reference lack a common measure. Finite-difference methods have also been applied to the study of the zeta function by leading specialists. But to make things even more interesting, we should remember that both methods do *not* always yield the same values.

This would simultaneously fulfill several major objectives. It would be possible to connect the simplest and most irreducible element of calculus with the most complex aspects. Something less appreciated, but no less important, is that it puts the idea of function in direct contact with that which is not functional —with that which does not change. Analysis is the study of rates of change, but change with respect to what? It should be with respect to what does not change. This turn is imperceptible but transcendental.

It should be remembered that Archimedes did not invent calculus, but he did invent *the problem* of calculus, by looking toward zero. Mathis is rectifying an approach that is now over 2,200 years old.

The direct method laughs in the face of the "computational paradigm" and its fervent operationalism. The now prevailing idea of intelligence as "predictive power" is reactive; in fact, it is no longer known whether it is the one we want to export to machines or the one humans want to import from them. Perceiving what does not change, even if it does not change anything, gives depth to our field. To perceive only what changes is not intelligence but confusion.

On the other hand, the definition of equilibrium conditions, as a sum and as a product, as an external and internal perspective, as applicable to closed and open systems, as intensive and extensive quantities with their possible breakdowns, as reversibility and irreversibility, are at the core of the algebraic aspects complementary to the analytical aspects that should contribute to make the intrinsic more explicit to get a deeper understanding of the zeta function.

If our view is somewhat correct, the cleaning and clarification of the enormous field of thermodynamics and entropy, with all the work that lies ahead, would be closely associated with the progress in understanding the zeta function. This does not seem a promising prospect, since it amounts to tons of dirty work far away from the exertions in the wondrous gardens of the theory of numbers. However, there is a hope that the connection between the two domains can be greatly smoothed out by analytical notions such as intrinsic entropy and the introduction of irreversibility in the foundation of analytical mechanics.

*

We have seen that the gauge fields of the standard theory, based on the invariance of the Lagrangian, clearly exhibit a feedback. They don't even hide it, and the theorists have been struggling with terms like self-energy and self-interaction for generations now, shooing them away like flies of their faces because they thought it was a "silly" idea. Now, the renormalization group at the base of the standard model and statistical physics, containing this self-interaction, is more or less the same that is currently used in the neural networks of deep learning and artificial intelligence —so something that the electromagnetic field already has built-in is being used externally. Let us call it a missed natural intelligence.

Nobody made the integers, and all else is the work of man. The zeta function is an ideal candidate for creating a collective network of emerging artificial intelligence around it. Why? Because in it, as with addition and multiplication, the external and internal aspects of the act of counting, what some would call the object and the subject, coincide. In this type of networks will be key to what extent the potential of the physical substrate is tapped, the connection between analog and digital components in a hybrid approach, the local and global equilibrium conditions, the algorithms to simulate the structure of the integers, the structures that try to perceive or filter them, the correlation spectrum, and a good number of aspects we cannot list here.

Thus it may be possible to create a collective emerging artificial intelligence alien to human intelligence defined by purpose, and yet with the possibility of communication with humans at different levels through a common act, counting, that, as purpose limits our scope, also for humans use to exist at very superficial levels. Intellection is a simple act and intelligence is a connection or contact of the complex with the simple, rather than the opposite.

As a prime example of an analytical totality indigestible for modern methods, the zeta function is already a great sign. Someone, in an Information Age typical turn, might say that, if any possible information is already in that function, we might as well look for the answers there instead of in the universe. But the opposite is true: to better understand the function we need to change the ideas of how the universe "works" and the range of application of our more general methods. That is the beauty of the subject.

Until now, the mathematical physics from which the entire scientific revolution emerged has been applying mathematical structures to physical problems in order to obtain a partial solution by means of an artful reverse engineering. Calculus itself emerged in this process. On the contrary, what we have here are physical processes that spontaneously reflect a mathematical reality for which there is no known solution. This already invites us to change our logic from top to bottom.

As for the information itself, if with Shannon the sequential aspect of the flow of information units prevails, as we move towards more massive amounts of data the relevance of correlations increases unstoppably, and they even tend to constitute an autonomous sphere.

The data we use directly is one thing and the metadata that can be elaborated with that data and its multiple correlations with

[179]

other data is quite another. The same can be said of a genetic sequence and the multifactorial analysis of the relationships between different genes, and so on. A Riemann sphere, or a Riemann surface with many layers like an onion, seems to be much more suitable representations for the multidimensional reality of analysis than the increasingly insignificant relationship of causal sequences.

However, the zeta function contains the most basic relationship between the simplest sequence, that of integers, and the infinite world of correlations between its elements. But we should not forget the most primary aspect of the question, on the contrary. We should not forget where problems come from, neither in information theory, nor in calculus, nor in mechanics.

In fact nearly everyone thinks that causation is something of the past; but this way we forget that even mechanical causation is a global issue, not a local one, and this may be relevant where we least expect it. If the unexplained dynamics of the zeta is not consistent with the foundations of mechanics, let us change the rules of mechanics, for the understanding of the problem deserves it. It is Nature that sings through random matrices and many other instances, not the spirit of the laws.

Today, it is no longer a question of mastering nature but of mastering technology, since the latter now has more destructive potential for human beings than the former. However, mastering technology demands freeing Nature from constraints imposed by a technoscience that is too instrumentalised from the outset. Considering the history of science as a whole, it is perhaps not so surprising that the avenger is now the most selfless of all sciences, the useless but eternal theory of numbers.

*

In a previous chapter we said that the concept of order is not less subjective than that of harmony; it goes without saying that this could be the subject of endless discussions. There are a lot of formal definitions of order in mathematics for as many different cases, but it could hardly be argued that the basis of them all are the natural numbers. In fact, without them mathematics would not be able to order anything.

The paradoxical aspects of entropy are related to the relative nature of the notion of order. Something with a highly visible order has more potential for disorder than what is already in complete disorder, which can be translated into the social paradox of entropy: the more complex the society, the more disorder it seems to produce. On the other hand, we also saw earlier that entropy is not necessarily the best measure of complexity, and that the energy rate density can be more revealing. These are far-reaching issues that should be attended carefully.

From another angle one can say that pure randomness is the ultimate order, something suggested  by the very arrangement of the primes within the number system, so chaotic at a local level and with such a striking global structure —or as one mathematician put it, growing like weeds and yet marching like an army.

The zeta function is a transformation of the traditional harmonic series $(1+1/2+1/3+1/4...)$ so that it does not diverge. The harmonic series, attributed to the Pythagoreans and also known in China at roughly the same time, gives us the harmonics or overtones added to the fundamental wavelength of a vibrating string, and it is a path to understand musical intervals, scales, tuning and timbre.

It has been precisely within music theory that Paul Erlich has combined information entropy with the theory of harmony to define a relative *harmonic entropy*. Harmonic entropy, "the simplest model of consonance... ask the question of how confused is my brain when it hears an interval. It assumes only one parameter in answering this question" [81].

Erlich's concept of harmonic entropy deepens the line of research opened by engineer and psychoacoustic Ernst Terhardt, who had already introduced notions such as virtual tone. Virtual tones, in contrast to spectral tones, are those that the brain extracts even if the signal is masked by other sounds. The ear has a strong propensity to adjust what it hears in one or a few harmonic series.

[181]

Harmonic entropy can be considered as a kind of "cognitive dissonance", but is also related to the intrinsic uncertainty of time series.

Harmonic entropy has a solid theoretical basis that makes extensive use of the Riemann zeta function. This closes a circle mostly unbeknownst, since Riemann himself made a memorable, though unfinished, study of the mechanism of the ear that already considered the global aspects of auditory perception and was conceptually much more advanced than Helmholtz's reductionist model that was his starting point.

Riemann was struck, among many other things, by the incredible sensitivity of the detectors in the inner ear, currently known to be able to perceive displacements smaller than the size of an atom, or 1/10 of a hydrogen molecule. Would it be possible to illustrate the behavior of the zeta function with a precise acoustic analogy? Can we bring it to that area where understanding and perception merge?

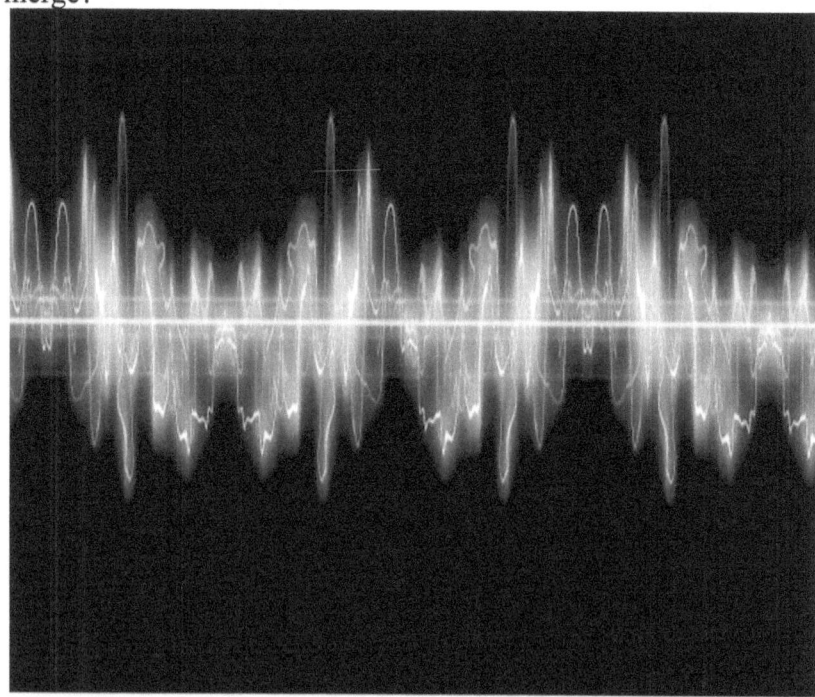

Certainly, issues related to the senses are not among the most important for mathematicians; but here it is easy to see that we are dealing with a problem of high interest at different levels, for laymen and experts alike. If deep math reaches our perception, just as deeply it changes our perception of math and numbers, and that has never been more necessary than now. There is a wide range of psychoacoustic experiments to survey the matter. Erlich uses Farey's series, one of the simplest ways to illustrate Riemann hypothesis.

Well known physicists have shown interest in the conversion of the zeta function and prime numbers into sound in order to hear its "music"; however the experiments we are talking about do not have to refer directly to this function, but first of all to the more generic questions of internal complementarity between spectral and virtual tones. The limits for the interest of the sonification of the arithmetic aspects would depend on the extent of this complementarity.

The reflection of the zeta function in perception in terms of harmonic entropy is a quest in its own right, and it should be sought with as much or even more eagerness than the physical systems capable of replicating it, as it is a more unitary endeavor. But in fact, they are not entirely different problems, even though they appear to us as the subjective and objective ends of the question, which is still misleading.

From the more utilitarian point of view of neural networks and artificial intelligence, the view that there is a basic continuity between perception and the so-called "higher" cognitive aspects is increasingly accepted. Riemann's ideas about hearing as an analytical abstraction are precisely those applied now, 150 years later, to the computer models that attempt to reproduce it.

Distributed networks such as the one mentioned above, for example, would lack something essential without an ability to filter or perceive numbers analogous to that of the ear and the brain when selecting and reinterpreting acoustic signals. Riemann's unfinished study of the ear, which calls the analogy "the poetry of hypothesis", dates from 1866. Weber-Fechner's psychophysical, differential and

[183]

logarithmic law, dates from 1860; Helmholtz's great work on auditory physiology, from 1863. In 1865 Clausius makes public the first definition of entropy.

No doubt Riemann was after something deep and relevant as usual for him —although he would have had to wait at least a hundred years to start putting the pieces together. Today we can do so, although not without first making some simple, but fundamental adjustments.

Nor is there any doubt that hearing has physical, but also cognitive and psychophysical limits, and that the latter act on the former in such a way that we are faced with variable thresholds defined by harmonic entropy. Within this interaction we would find the zeta function.

Is there a place for the mathematics of harmony, based on the continuous proportion, within this context of the harmonic series and its adventures in the unlimited space of analysis? Certainly not, if one thinks of explicit arithmetic relations; at least so far mathematicians have not found anything worth mentioning. This would lead us back to the hypothesis that from Euclid's *Elements,* unmixed like water and oil, two different traditions flow.

In the world of appearances, water and fire do not come into contact except in the guise of a cloud or mist of vapor. Since the first chapter we have been wondering what the relationship of the continuous proportion to analysis might be and we have not yet a definitive answer on the subject, but one of the threads would certainly be the connections of this proportion with entropy, elementary calculus and the algorithmic theory of measurement.

Needless to say, the difference between algebraic and transcendental numbers, so important in analysis and arithmetic, is irrelevant to acoustics and most of practical problems.

Thus, the harmonic series and analysis, the world of waves that reflects forms, would be on the "water shore" of reversibility; the continuous proportion, with its capacity to combine the continuous and the discrete, on the "fire shore" of entropy as irreversibility.

Of course, analysis and synthesis presuppose each other, but there are uncountable levels for their interaction. The zeta function, an analytical totality with a single pole, is in itself a prodigious synthesis, which also lends itself to all kinds of transformations in alternating, symmetrical functions, and so on. Now, the main synthetic component in modern science is hidden in the very idea of the elements, as fundamental building blocks —in physics atoms and particles. On account of them, it was thought unnecessary to look for other constructive and synthetic resources.

This is a burden from the past of physical atomism that information theory has yet to overcome, since the idea of independent elements will always be less restrictive than the net of observable, empirical correlations.

We will leave this topic suspended in the mist of its own cloud, trying to recall what is beyond complementarity, analysis and synthesis, object and subject. The example of the informal calculus of the outfielder after the the fly ball tells us that both the ball and the runner move mutually and correlatively with respect to what does not move —the constant mean.

This is the true invisible axis of activity, since the thinker is not less a thought than any other. What else can be said? But winged thought always flees from the zone of stillness, that is precisely what its life consists of. Surely this is too simple for us, and that is why we have discovered and invented the harmonic series, the continuous proportion, the entropy or the zeta function.

*

Bernhard Riemann's eight-page paper *On the number of primes less than a given magnitude* was published in November 1859, the same month as *The origin of species*. That very same year saw the beginning of statistical mechanics with a pioneer work by Maxwell, and the starting point of quantum mechanics with spectroscopy and the definition of the black body by Kirchhoff.

I have always felt that in Riemann's conscientious study, so free of further purpose, there is more potential than in all of quantum mechanics, statistical mechanics and its offspring information theory, and the theory of evolution combined; or that at least, in a secret balance sheet, it constitutes a fair counterweight to all of them. Although if it helps to redirect the doomed trend of infonegation of reality it will have already been enough.

These three or four developments mentioned are after all children of their time, circumstantial and more or less opportunistic theories. It is true that Boltzmann fought and despaired for the assumptions of the atomic theory to be admitted, but he had his whole battle with physicists, because chemists had already been working with molecules and atoms for a while. In mathematics, however, the lags operate on another scale.

A wordy bestseller, *The origin of species* was discussed in cafés and taverns on the very day of its presentation; but the Riemann hypothesis, being just the strongest version of the prime number theorem, nobody knows how to go about it after 160 years. Clearly we are talking about different wavelengths.

And what does one text have to do with the other? Nothing at all, but they still present an elusive point of contact and extreme contrast. The usual reading of the theory of evolution says that what drives the appreciable order is chance. Riemann's hypothesis, that prime numbers are distributed as randomly as possible, but that randomness hides in itself a structure of infinite richness.

Mathematicians, even living on their own planet, are gradually realizing that the consequences of proving or disproving Riemann's hypothesis could be enormous, inconceivable. It is not something that happens every Sunday, or every millennium. But we do not need to ask so much: it would be enough to properly understand the problem for the consequences to be enormous, inconceivable.

And the inconceivable becomes more conceivable precisely as we enter the Information Age. The inconceivable could be something that simultaneously affects our idea of information, its calculation and its physical support. Both software and hardware: a full blown short-circuit.

Probably information is not destiny, but the zeta function may be the destiny for the Information Age, its final event horizon. This would take us away from the spectre of the "technological singularity" and bring us closer to a very different landscape. The zeta function has a single pole that is dual with the zeros —the zeros reflect information that the unit cannot give because there are no finite results for s =1. This suggests an enigmatic envelope of reflectivity for the information galaxy.

In all probability, the substance of the problem is not a technical issue. But many of its consequences are, and not only technical. This development has largely gone against the grain of other scientific developments, and its assimilation would profoundly affect the dynamics of the system as a whole.

Like the prime numbers themselves, in the chronology and the sequential order the events cannot seem more fortuitous, but from another point of view they seem destined by the whole, pushed by the ebb from all shores to fit in the very moment they took place. Riemann lived in Germany in the middle of Hegel's century, but in its second half the opposition to idealism could not have been greater in all areas, and in the sciences in particular. The pendulum had turned with all its force and the sixth decade of the century marked the peak of materialism.

Also a son of his time, Riemann could not help but feel acutely the contradictions of 19th-century liberal materialism; but the German mathematician, a theoretical and experimental physicist close to Weber and a profound natural philosopher, was an heir of Leibniz and Euler continuing their legacy by other means. His basic conviction was that we humans do not have access to the infinitely large, but at least we can approach it through the study of its counterpart, the infinitely small.

These were prodigiously fruitful years for physics and mathematics. Riemann passed away at thirty nine and did not have time to articulate a synthesis in tune with his conceptions and permanent search for unity; a synthesis understood not as an arbitrary construction, but as the unveiling of the indivisibility of truth. But he achieved something of the kind where least expected: in the analytical theory of numbers. A provisional synthesis though, that has given rise to the most famous conditional of mathematics: "If the Riemann's hypothesis is true, then…"

This unexpected and conditional synthesis came about through complex analysis, just in the same years that complex numbers were beginning to emerge in physics, seeking, like atoms and molecules, more degrees of freedom, more room to move. The role of complex numbers in physics is a topic always postponed since it is assumed that their only reason for being is convenience — however, when it comes to dealing with the zeta function and its relationship with physics, there is no mathematician who is not forced to interpret this question one way or another, which physicists generally associate with rotations and amplitudes.

But complex analysis is just the extension of real analysis, and to get to the heart of the matter we must look further back. Riemann's conditional synthesis speaks to us of something indivisible, but it still relies on the logic of the infinitely divisible; not to solve the famous problem, but simply to be in tune with it, it would have to be understood in terms of the indivisible itself, whose touchstone is the constant differential.

Is there anything beyond information and the computer? Of course there is, the same thing that is waiting for us beyond the Symplegades. A much more vast and undivided reality.


## (Added June 10, 2022)

Scot C. Nelson [82] discovered in late 2001 that the logarithmic spirals of plant growth —sunflowers, daisies, pine cones, etc. — "serve as a simple and naturally efficient prime number sieve". Like everything related to the continuous proportion and its associated number series, this has received barely any attention and seems relegated in advance to the ever-growing section of anecdotal coincidences. And yet this is the first basic connection that has been found between prime numbers and these ubiquitous patterns of phyllotaxis, which should have told us something. In light of Nelson's finding, there appears to be a "central symmetry of prime numbers within three-dimensional objects", and vegetal growth would have natural prime number generating algorithm in its becoming. The same passage from the number line to the unfolding of these patterns on surfaces and in three dimensions should be a thread for the geometric intuition of the fundamental theme of arithmetic. It is not the same to try to link arithmetic with modern abstract "geometry" than to link it with a natural geometry. A mechanical analogy comes to mind: the parts repel each other like magnetic dipoles with a minimization of the energy between them, and as the plant grows the time delay between the formation of new primordia is reduced. One can think of it in ergontropic and information entropy terms as well.

# Referencias

[1] John Arioni, *Golden Ratio in Yin-Yang*
https://www.cut-the-knot.org/do_you_know/GoldenRatioInYinYang.shtml

[2] Alexey Stakhov; *The Mathematics of Harmony — From Euclid to Contemporary Mathematics and Computer Science*
http://vixra.org/pdf/1602.0042v1.pdf
Ervin Wilson, *The scales from the slopes of Mt. Meru and other recurrent sequences,*
http://anaphoria.com/wilsonmeru.html

[3] Agno, *Imaginary Golden Ratio, My math forum*
http://mymathforum.com/number-theory/17605-imaginary-golden-ratio.html

[4] Math *Train, Imaginary Golden Ratio, StackExchange*
https://math.stackexchange.com/questions/1851698/imaginary-golden-ratio
Karl Dilcher, *Hypergeometric functions and Fibonacci numbers,*
https://www.fq.math.ca/Scanned/38-4/dilcher.pdf
*Rogers-Ramanujan continued fraction*
https://en.wikipedia.org/wiki/Rogers%E2%80%93Ramanujan_continued_fraction

[5] René Guenon, *Symbolism of the Cross,* Chapters XX-XXII.

[6] Peter Alexander *Venis, Infinity-theory —The great Puzzle*
http://www.infinity-theory.com/en/science/Main_pages/The_Great_Puzzle

[7] Richard Merrick, *Harmonically Guided Evolution*
http://interferencetheory.com/files/Harmonic_Evolution.pdf

[8] A. K. T. Assis, *The principle of physical proportions*
*Relational Mechanics and Implementation of Mach's Principle with Weber's Gravitational Force* (Apeiron, Montreal, 2014), 542 pages, ISBN: 9780992045630
[9] Nikolay Noskov, http://bourabai.kz/noskov/index.html
*The Phenomenon of Retarded Potentials,*
*The Theory of Retarded Potentials,*
*Limits of application of fields in classical mechanics*

[10] Miles Mathis, http://milesmathis.com/
*The Physics behind the Golden Ratio*

More on the Golden Ratio and Fibonacci Series

[11] Miles Mathis, Explaining the Ellipse
*Unlocking the Lagrangian*

[12] Nicolae Mazilu, *A Newtonian Message for Quantization*
Nicolae Mazilu & Maricel Agop, *The Mathematical Principles of the Scale Relativity Physics I. History and Physics*

[13] Roger Tattersall, *Why Phi? Simplified: A brief Fibonacci tour of the Solar System*
Jan Boeyens, *Commensurability in the solar system*
Harmut Müller, *Global Scaling of Planetary Systems*

[14] Miles Mathis, *A Complete Correction to and Explanation of Bode's Law. The average deviation Mathis finds, based on a simple orbital law, is 2.75 %, exactly the same Roger Tattersall gives for the golden spiral.*
Miles Mathis, *A Redefinition of Gravity. Part IX. Why the Sun and Moon have the same Optical Size*

[15] Mario J. Pinheiro, *A reformulation of mechanics and electrodynamics*

[16] Miguel Iradier, *La Tecnociencia y el Laboratorio del Yo*
Rod Swenson, M. T. Turvey, *Thermodynamic Reasons for Perception-Action Cycles*

[17] Richard Merrick, *Harmonic formation helps explain why phi pervades the solar system*

[18] Gian Paolo Beretta, *What is Quantum Thermodynamics?*

[19] Xue-Jun Zhang and Zhong-Can Ou-Yang, *The Mechanism Behind Beauty: Golden Ratio Appears in Red Blood Cell Shape,*
Marcy C. Purnell and Risa D. Ramsey (February 7th 2019). *The Influence of the Golden Ratio on the Erythrocyte*

[20] Miles Mathis, *Perturbation Theory in the Light of Charge*

[21] S.C. Tiwari, *Geometric Phase in Optics and Angular Momentum of Light*
Alexander Ershkovich, *Electromagnetic potentials and Aharonov-Bohm effect*

[22] A. K. T. Assis, *Relational Mechanics and Implem*entation of Mach's Principle with Weber's Gravitational Force (Apeiron, Montreal, 2014), 542 pages, ISBN: 9780992045630

[23] Alejandro Torassa, *On classical mechanics,*

[24] René Guenon, *The metaphysical principles of the infinitesimal calculus, chapter XVII, Representation of the balance of forces*

[25] A. K. T. Assis, *History of the 2.7 K Temperature Prior to Penzias and Wilson*

[26] Stephen Spurrier, Nigel R. Cooper *Semiclassical Dynamics, Berry Curvature and Spiral Holonomy in Optical Quasicrystals*

[27] Nicolae Mazilu, *Mechanical Problem of Ether*

[28] C.K. Thornhill, *The foundations of relativity*

[29] Patrick Cornille, *Replication of the Trouton-Noble Experiment*
*Patrick Cornille, Simple electrostatic aether drift sensors (SEADS): New dimensions in space weather and their possible consequences on passive field propulsion systems*

[30] Yuchao Zhang, Jie Gao, and Xiaodong Yang, *Optical Vortex Transmutation with Geometric Metasurfaces of Rotational Symmetry Breaking,*

[31] Michael Berry, *Quantal phase factors accompanying adiabatic changes Geometric phases*, https://michaelberryphysics.files.wordpress.com/2018/03/berryd.pdf

[32] John Carlos Báez, *Black Holes and the Golden Ratio*

[33] Yasuichi Horibe, *An entropy view of Fibonacci Trees*
*Takashi Aures, The Fibonacci sequence in nature implies thermodynamic maximum entropy*

[34] Miguel Iradier, *Towards a science of health? Biophysics and Biomechanics*

[35] Miguel Iradier, *Beyond control: feedback and potential*

[36] Mario J. Pinheiro, *A Variational Method in Out-of-Equilibrium Physical Systems*
Patrick Cornille, *Simple electrostatic aether drift sensors (SEADS): New dimensions in space weather and their possible consequences on passive field propulsion systems*

[37] R. Ferrer i Cancho, A. Hernández Fernández, *Power laws and the golden number*

[38] Aram Z. Mekjian, *Power law behavior associated with a Fibonacci-Lucas model and generalized statistical models*
Ilija Tanackov, *The golden ratio in probabilistic and artificial intelligence*
Myron Hlynka and Tolulope Sajobi, *A Markov chain Fibonacci Model*

[39] Soroko. E.M., *Structural Harmony of Systems*. Minsk: Nauka i Technika (1984) (Ruso). Comentado por A. Stakhov en [2].

[40] Yaniv Dover, *A short account of a connection of Power Laws to the Information Entropy,*
Matt Visser, *Zipf's law, power laws and maximum entropy*

[41] Mitchell Newberry, *Self-Similar Processes Follow a Power Law in Discrete Logarithmic Space*

[42] G. Rotundo, M. Ausloos, *Complex-valued information entropy measure for networks with directed links* (digraphs)

[43] A. Stakhov, op. cit. Petoukhov, S.V. *Metaphysical aspects of the matrix analysis of genetic code and the golden section*. Metaphysics: Century XXI. Moscow: BINOM (2006), 216250 (Ruso)

[44] Fernando C. Pérez-Cárdenas, Lorenzo Resca, Ian L. Pegg, *Coarse Graining, Nonmaximal Entropy,*
*and Power Laws*

[45] Miles Mathis, *A Re-definition of the Derivative (why the calculus works— and why it doesn't)*
Miles Mathis, *Calculus simplified*
Miles Mathis, *My calculus applied to exponential functions*
Harlan J. Brothers, John A. Knox, *New closed-form approximations to the Logarithmic Constant e,*

[194]

John A. Knox, Harlan J. Brothers, *Novel Series-based Approximations to e*

[46] Paul Marmet, *The Subjectivity of Heisenberg's Uncertainty Relationship*
Miles Mathis, *One Thing is Certain: Heisenberg's Uncertainty Principle is Dead*
Wikipedia, *Intensive and extensive properties*
V.V. Aristov, *On the relational statistical space-time concept*
V. V. Aristov. *Relative Statistical Model of Clocks and Physical Properties of Time*
S E Shnoll, V A Kolombet, E V Pozharski⌐, T A Zenchenko, I M Zvereva, A A Konradov, *Realization of discrete states during fluctuations in macroscopic processes*
[47] Peter Alexander Venis, *Infinity theory*

[48] Michael Howell, *Logarithmic derivatives sans the diminishing differentials*
http://milesmathis.com/howell4c.pdf

[49] Igor Podlubny, *Geometric and Physical Interpretation of Fractional Integration and Fractional Differentiation*

[50] Miles Mathis, *Electrical Charge*

[51] Evert Jan Post, *Quantum reprogramming —A long Overdue and Least Intrusive Reality Adaptation of the Copenhagen Interpretation*

[52] Dániel Schumayer, David A. V. Hutchinson, *Physics of the Riemann Hypothesis*

[53] Nicolae Mazilu, *Physical Principles in Revealing the Working Mechanisms of Brain, Part One*
Nicolae Mazilu, *The Classical Theory of Light Colors: a Paradigm for Description of Particle Interactions*

[54] Nicolae Mazilu, *A case against the First Quantization* (2010)

[55] C. K. Thornhill, *The kinetic theory of electromagnetic radiation*

[56] Irwin G. Priest, *A Proposed Scale for Use in Specifying the Chromaticity of Incandescent Illuminants and Various Phases of Daylight, (1932)*

[57] Paul Marmet, op. cit.

[195]

[58] David Hestenes, *Quantum Mechanics from Self-Interaction*
P. Catillon · N. Cue · M.J. Gaillard · R. Genre · M. Gouanère · R.G.
Kirsch · J.-C. Poizat · J. Remillieux · L. Roussel · M. Spighel, *A Search for the de Broglie Particle Internal Clock by Means of Electron Channeling*
Shau-Yu Lan, Pei-Chen Kuan, Brian Estey, Damon English, Justin M.
Brown, Michael A. Hohensee, Holger Müller, *A Clock Directly Linking Time to a Particle's Mass*, Science 339, 554 (2013)
George Savvidy and Konstantin Savvidy, *Quantum-Mechanical Interpretation of Riemann Zeta Function Zeros*
Timothy H. Boyer, *Stochastic Electrodynamics: The Closest Classical Approximation to Quantum Theory,*
Nicolae Mazilu, Maricel Agop, *Role of surface gauging in extended particle interactions: The case for spin*

[59] Mario J. Pinheiro, *On Newton's Third Law and its Symmetry-Breaking Effects*
Koichiro Matsuno, *Information: Resurrection of the Cartesian Physics*

[60] E. J. Chaisson, *Energy Rate Density as a Complexity Metric and Evolutionary Driver,*
*Energy Rate Density. II. Probing Further a New Complexity Metric,*

[61] Georgi Yordanov Georgiev, Erin Gombos, Timothy Bates, Kaitlin Henry, Alexander Casey, Michael Daly, *Free Energy Rate Density and Self-organization in Complex Systems*

[62] A. Stakhov, op. cit. pgs. 120-125

[63] Vladimir A. Lefebvre:
https://www.researchgate.net/scientific-contributions/2034094747_Vladimir_A_Lefebvre
*The algebra of conscience* (review by J. T. Townsend),
https://www.researchgate.net/publication/256246817_Algebra_of_conscience_Vladimir_A_Lefebvre_Boston_Mass_Reidel_1982

[64] Raymond Abellio, *La structure absolue, Essai de phénoménologie génétique*, París, 1965

[65] McBeath, M. K., Shaffer, D. M., & Kaiser, M. K. (1995) *How baseball outfielders determine where to run to catch fly balls*, Science, 268(5210), 569-573.

[66] Víctor Gómez Pin, Filosofía. *El saber del esclavo*. Editorial Anagrama, 1989.

[67] Miles Mathis, *A Disproof of Newton's Fundamental Lemmae*
*A correction to the equation a = v2/r* (and a Refutation of Newton's Lemmae VI, VII & VIII)
*What is pi?*
*The extinction of pi*
*The Manhattan Metric*
*The Cycloid and the Kinematic Circumference*

[68] Gopi Krishna Vijaya, *Calculus and Geometry*
*Celestial Dynamics and Rotational Forces In Circular and Elliptical Motions*
*Original Form of Kepler's Third Law and its Misapplication in Propositions XXXII-XXXVII in Newton's Principia (Book I)*

[69] K. K. Nambiar, *Information-theoretic equivalent of Riemann Hypothesis (2003).*
J. R. Higgins, *The Riemann Zeta Function and the Sampling Theorem (2009)*
Er'el Granot, *Derivation of Euler's Formula and ζ(2k) Using the Nyquist-Shannon Sampling Theorem (2019)*

[70] K. K. Nambiar, *Electrical equivalent of Riemann Hypothesis (2003)*
Guðlaugur Kristinn Óttarsson, *A ladder thermoelectric parallelepiped generator (2002)*
Danilo Merlini, *The Riemann Magneton of the Primes (2004)*
M. V. Berry, *Riemann zeros in radiation patterns: II.Fourier transforms of zeta (2015)*
B. Van der Pol, *An electro-mechanical investigation of the Riemann zeta function in the critical strip (1947)*

[71] Matthew Watkins, *Number Theory and Entropy; Number Theory and Physics Archive*

[72] Rod Swenson, M. T. Turvey, *Thermodynamic Reasons for Perception-Action Cycles*
[73] Luciano Floridi, *What is the Philosophy of information?*

[74] O. Shanker, *Entropy of Riemann zeta zero sequence (2013)*
Alec Misra, *Entropy and Prime Number Distribution; (a Non-heuristic Approach) (2006)*

[197]

[75] Piergiulio Tempesta, *Group entropies, correlation laws and zeta functions (2011)*

[76] Gian Paolo Beretta, *What is Quantum Thermodynamics (2007)* También puede visitarse la página http://www.quantumthermodynamics.org/

[77] Miguel Iradier, *La estrategia del dedo meñique*

[78] Ian Wright, *Notes on a Hegelian interpretation of Riemann's Zeta function (2019)*

[79] Ryusuke Hamazaki, Kazuya Fujimoto, and Masahito Ueda, *Operator Noncommutativity and Irreversibility in Quantum Chaos (2018)*

[80] Ali H. Chamseddine, Alain Connes and Walter D. van Suijlekom, *Entropy and the spectral action (2018)*

[81] *Harmonic Entropy, Xenharmonic Wiki*
*The Riemann Zeta Function and Tuning, Xenharmonic Wiki*
Paul Erlich, *On Harmonic Entropy*, con comentario de Joe Monzo

[82] Scot. C. Nelson, *A Fibonacci Phyllotaxis Prime Number Sieve (20*